Of the Hunter Class

David Griffin

To Hugh ex S.A.S.

From a fellow member.

Best wishes & good luck.

David Griffin

R. Mc

Note for Librarians: a cataloguing record for this book that includes Dewey Decimal Classification and US Library of Congress numbers is available from the Library and Archives of Canada. The complete cataloguing record can be obtained from their online database at:
www.collectionscanada.ca/amicus/index-e.html
ISBN 1-4120-5463-X

Printed in Victoria, BC, Canada

Printed on paper with minimum 30% recycled fibre. Trafford's print shop runs on "green energy" from solar, wind and other environmentally-friendly power sources.

TRAFFORD Offices in Canada, USA, Ireland and UK

This book was published on-demand in cooperation with Trafford Publishing. On-demand publishing is a unique process and service of making a book available for retail sale to the public taking advantage of on-demand manufacturing and Internet marketing. On-demand publishing includes promotions, retail sales, manufacturing, order fulfilment, accounting and collecting royalties on behalf of the author.

Book sales for North America and international:
Trafford Publishing, 6E–2333 Government St.,
Victoria, BC V8T 4P4 CANADA
phone 250 383 6864 (toll-free 1 888 232 4444)
fax 250 383 6804; email to orders@trafford.com

Book sales in Europe:
Trafford Publishing (UK) Ltd., Enterprise House, Wistaston Road Business Centre,
Wistaston Road, Crewe, Cheshire CW2 7RP UNITED KINGDOM
phone 01270 251 396 (local rate 0845 230 9601)
facsimile 01270 254 983; orders.uk@trafford.com
Order online at:
trafford.com/05-0361

10 9 8 7 6 5 4

"IT IS UPON THE NAVY UNDER
THE GOOD PROVIDENCE OF
GOD, THAT THE WEALTH,
SAFETY AND STRENGTH OF
THE KINGDOM DO CHIEFLY
DEPEND."

Unknown*

*See back page author's note on the misconstruction of history.

This Book

is

dedicated

to

John Griffin

"And the man became legend"

CONTENTS

CHAPTERS

FOREWORD

Most critics in the book trade will always warn you, to catch and hold a reader of any calibre you must hook his attention span and then hold it somewhere within the first half page of your story. If you fail to deliver in this brief period, there is every chance he will continue to flick through the pages and then drop the book back on the shelf and to him, your story will never be told. In the area of worthwhile history, this failure of the written word is a shame and waste. For good or bad it is a harsh reality that today more and more people are finding a rare fascination with the military, and especially military history. But the area that stands out above all the masses of military might at hand today are the elite specialist elements and this is why I want to tell you a story in critical detail of the Royal Marines – and indeed the Royal Navy.

You can name 70% of all army regiments now in service and the public in general will have no idea of who you are talking about or brigades you are referring to. Well, I must tell you in my view this is because on saying goodbye to National Service away back in 1962, it took just one generation to lose the public pride that once existed within the communities throughout the country for their own county regiments. To most people today, soldiers are just "The Army"! I fear Rudyard Kipling (to whom we will on numerous occasions refer for wisdom during our story) may have been right when his poetry skirted the peripheries of public indifference for "Tommy Atkins".

"For it's Tommy this, an' Tommy that, an' 'chuck him out, the brute!'

But it's saviour of 'is country when the guns begin to shoot.'"

How true! Now, what is my point? My point here is three household words – 'The Royal Marines'. I want you to forget about the enormity of American – Jap – and German divisions hurtling against each other across miles of battlefield terrain, because the warriors I have in mind are of a much more limited but specialist nature. To tell the real story in very fine and personal detail, I fear the dreamed-up adventure stories of military this and that you can find in most High Street shops today were of no use whatsoever to me, because most are, in some areas, perforated with inaccuracies and fanciful doers of great deeds.

To compose "Of the Hunter Class", I had to search and wipe the dust, as it were, from the archives of the original records of what was the "Ministry of Information", marked as His Majesty's Stationery Office, London, 1944. I searched weeks into months in the magnificently co-operative libraries of the Imperial War Museum, The Royal Marines Museum, The Royal Naval College, Greenwich and many other private sources who have asked not to be mentioned for family and personal reasons, which I have observed and respected. All other names, events, naval engagements and land skirmishes brought to light are on record at the Ministry of Information.

"Of the Hunter Class" is not a huge novel but a true history of when, where, how and why the Royal Marines Commandos as we know them today came about; the rivals, the impostors, the training failures, but primarily their spectacular success that grasped and held the imagination of the world, especially that of the lesser military.

If you will, I would like to take you on a very exhaustive and uncomfortable journey, but more so I want to prove to you that nothing in life which is really worthwhile, admired and emulated right across the globe is easy.

We demand from these men a monumental contribution in effort, common sense, forward thinking, sheer guts and determination and a will to win which confounds all our enemies in all theatres of operation.

At this point I think it fair to remind you, if you are still reading then you must be on that 'hook' we mentioned at the beginning, and if you are you will agree when I say the ones who count most here are those who make the history and not the likes of us who sit with pen and parchment to tell the great tale. It is the sweat, the pain, the discipline and doing that counts.

From the mountainous heights of Robins Nest, which straddles the borders of Hong Kong and China, across the globe to the boiling pits of the Devil's Anvil north of Aden in the Yemen, and on to the drenched, steaming jungles of Malaya, I have crossed these theatres with most elements of the Marines, but where precisely the commando organisations were dreamed up and forged is to most people a mystery, even to the modern Army.

On these travels, I would like to mention the dark, brooding church spires of South Armagh, but there is a point in Irish history which has not yet been made, for I firmly believe the endgame in the island of Ireland has yet to cross the bloodiest of terrain in the theatre of paramilitarism and civil tragedy in which the Royal Marines will yet again be forced to play the professional mediator. I have said this in "The Fourth Province", but as in so many areas of world conflict the unwillingness to observe the facts at 'street level' – for it is the people who hold the key - is the favoured blind ploy of the 'here today and gone tomorrow' politicians. They have yet to learn that the price of faltering indifference can be catastrophic.

But let us leave Ireland for another day, because a true hunting class journey is imminent. It all started because of a similar lack of forethought at political level to specialist military needs, which burdened Imperial Britain with a huge problem. I think this was the fault of people at various levels of power who harboured a simplistic lack of understanding to the needs and logistics of specialist operations.

Let me now keep you on the reading hook of military history, which I believe is on the increase as a result of the passage of time itself. We are at a time when those who took part in the making of World War Two history are disappearing at an alarming rate, and we are now seeing an upturn in post-National Service and very many more younger readers developing a keen interest in how it actually was. So to 'paraphrase' the Marines – "Saddle up" – and come back in time with me to a dull, cold, wet and utterly depressive late morning in September of 1939.

David Griffin
Ringwood, Hampshire
January 2003

MAKING A MARINE

The contribution of the Royal Marines to British Defence towards the latter end of the twentieth century became legendary from Borneo to the Arabian Peninsula, from re-taking the Falkland Islands to peacekeeping/policing the streets of Northern Ireland. Wherever you care to mention, behind each of the troops wearing the Green Beret were some of the toughest selection and training courses in the entire military orientated world. I have known squads of sixty or more recruits on selection pared down to twenty-five or even less. One instructor once put it to me it was akin to panning for gold, when you wash and sift through the rough gravel, the shining residue left is the calibre of metal you were searching for.

When selection is over and the few successful candidates are informed they are going ahead to full training, I am sure the same thoughts drift through each of their minds; 'God ha' mercy on such as we.' I know it did through mine.

After the Falklands War, the Ministry of Information described officer training in many variations for general publication and of course for recruiting purposes, but for me the recruiting poster should also have said that to become a Royal Marine you have to be able to do everything the British Army does, only better and faster. Since the end of World War II, when the Army Commandos were disbanded, Britain's Commando forces have been provided by the Royal Marines. The Royal Marine Commandos distinguished themselves during that war, and have continued to do so in many small and limited wars around the globe. Becoming a commando, as mentioned above, is part of the selection and training process when joining the corps of the Royal Marines.

Ever since that epic "yomp" across East Falkland Island from the beach head at San Carlos Sound to the hills overlooking Port Stanley, the stamina and fitness of the Royal Marines has caught the public's imagination. However, the way of becoming a Commando remains something of a closed book to most people.

It is not easy to win the Green Beret, but it is worth the effort when you do. The initial days and weeks are the worst. There is no let-up; exercises in all weathers. You wish at times you had never begun. It gets easier though, as you get fitter. In physical terms, the winning of the Green Beret means a lot of sweat in the gymnasium, on the parade ground and in the field. Out of doors, the would-be Marine has to run with rifle and personal load-carrying equipment (C.E.F.O.), climb, vault, swim and clear obstacles. There are cross-country marches by day and by night, and long distance walks in set times with aching limbs, and boots seem to get heavier and heavier. As training increases in pace and intensity, the recruit realises that he is far short of the required standards. But the troop instructors ensure that he improves on a daily basis, and to his own great surprise he finds that he actually does.

Together with the physical conditioning, the recruit – in common with all servicemen – learns the basics of the trade; to be a competent infantryman he learns to fire, strip, clean and master his personal weapon, and also the section fire and support weapon, usually the (G.P.M.G.) – General Purpose Machine Gun. There is also instruction on the other weapons commonly used in the infantry section.

He is lectured in the classroom, and applies this knowledge in practice periods in the field; on camouflage (which is often nowadays called by the more comprehensive term of 'counter-surveillance'), infantry minor tactics and survival techniques.

Mastering the Basics

The recruit learns military communications, both written and spoken, and how to work with helicopters. Helicopter emplaning and deplaning drills are a simple subject, but they must be mastered as all soldiers use them – and the Royal Marines probably more than most. The recruits go to the base at Poole in Dorset for a week, where they are introduced to seamanship and amphibious warfare techniques. They are taught about the ships in which the Royal Marines serve and they see the Special Boat Service, (S.B.S.), at work.

About halfway through the recruits' training, everything seems to come together for them. The recruit seems to be fitter, smarter and much more 'switched on' to most situations. He is more agile on ropes and confident of heights. He now begins to train in earnest for the coveted Green Beret. Skills learned are perfected, individual skills are applied to the needs of team tactics, and the other weapons which a Commando uses are learned and mastered. The recruit participates in section and troop tactical exercises in the field and in live firing of the weapons.

He also takes part in fire and movement with live ammunition, goes climbing and learns such back-up skills as first aid. During the field training exercises, defensive and assault positions are prepared, which can mean back-breaking digging with entrenching tools, depending on the terrain, and then the recruit applies the counter-surveillance lessons to give concealment from ground and airborne observation.

Towards the end of the Recruits' Training Course, the four-day long Commando tests are taken. The aim of these is to show what the recruit can accomplish and what personal resources he has. The tests involve, first, a nine-mile speed march with full equipment and rifle, which has to be completed in 90 minutes.

Second, there is the endurance course. This consists of a four-and-a-half mile run with rifle and equipment over bog and rough country, overcoming obstacles and crawling through tunnels – one of which is filled with water. This test has to be completed in 80 minutes, and finishes on a rifle range where marksmanship must be proved. At this point, it is worth remembering that normal target competition shooting is performed by relaxed competitors with calm deliberation, but the marksmanship demonstrated at the end of the endurance course has to be performed by men who are breathless, aching and exhausted, with wet filthy weapons covered in mud and grit, which are more often than not prone to jamming, and because it is a timed exercise a jammed weapon has to be cleared and fired in as little time as possible. As every minute counts, it goes without saying that this also becomes an exercise in nerve-rattling frustration.

Third is the assault course, which must be cleared in five minutes. This consists of crossing varying obstacles in the time, carrying rifle and equipment.

Fourth is what is loosely known as the 'Tarzan Course'. This is a series of rope bridges, well above the ground, which has to be crossed in five-and-a-half minutes, carrying rifle and equipment.

The fifth test is a demonstration of the would-be Commando's ability to handle all troop weapons by passing the appropriate handling tests. The sixth and last is the swimming test. It is generally held that the would-be Royal Marine officer has to pass all these in less time than the men. Having successfully passed, the recruit wins the Green Beret and becomes a Royal Marine.

Before all this happens, the would-be Commando has to be selected. True, during the 24-week long recruits' course, some are weeded out. But before applicants join this, there is a preliminary selection by a two-day Potential Recruits Course (P.R.C.). About 1,000 young men join the Royal Marines each year, and as many more again volunteer but, for one reason or another, as I mentioned at the beginning, fail to meet the entry standards.

'Earning the Beret'

If the normal requirements of age, nationality and physical fitness are met, the recruit is enlisted in either a career engagement or a notice engagement, and is sent to the Commando Training Centre (C.T.C.) at Lympstone in Devon. There he begins a 26-week Recruit Training Course. The first two weeks of this is really a sort of induction course, during which recruits are briefed on all aspects of the future job and are prepared for more intensive training. The recruit may be discharged – or discharge himself – during these two weeks if he fails to meet the required standards. One young officer who springs to mind during such periods of training was Prince Edward, after which the press had a field day, but which, in my opinion, was totally unwarranted, as the sheer intensity of Special Forces Training is not for everybody.

After this period, a 24-week 'Beret Course' begins. Officers' selection takes place in two parts. The first stage is the Potential Officers' Course (P.O.C.) which lasts two days and takes place at the C.T.C. Lympstone. This P.O.C. consists of a variety of tests and exercises, which gives the instructional staff the opportunity to evaluate the potential officers' physical and mental aptitude for leadership. The P.O. is watched carefully as he performs in the gymnasium, on an endurance run, and on the 'Tarzan Course'. There are also tests of spoken and written ability, such as an essay on current affairs topics and a brief lecture to a small audience for five minutes. If the P.O.

passes this course, he is recommended to go on to the second stage of the selection procedure, which is the Admiralty Interview Board.

The Admiralty Interview Board takes place 'on board' H.M.S. Sultan, which is a naval shore establishment at Gosport in Hampshire. It is chaired by a Rear Admiral and usually consists of five members including the President (a Royal Marines Colonel or Lt. Colonel), a Personnel Selection Officer (a Lt. Commander R.N.) and a civilian Headmaster or University tutor. This Board looks more deeply into the P.O.'s potential over the next day and a half of the A.I.B. Here it examines motivation, character, personality, initiative and ability to lead. ('Will we ever see our politicians put through such searching selective mills, I say to myself!')

The final decision rests with the Ministry of Defence in London, and depends on the report of the A.I.B. and then on the number of vacancies available. If the Potential Officer is successful, he is sent to the C.T.C. at Lympstone for a two-year training period, during which time he is given a probationary commission in the Royal Marines.

The first part of the course is aimed at passing the Commando Test and winning the Green Beret. It is tougher for the P.O. than for the men; the Officer, for example, has to complete the 30-mile speed march (and it is a fast pace) in seven hours, whereas the Marine has eight hours to do it in.

The second part consists of military training of a tactical nature. The second year is spent in the field as a Rifle Troop Commander. At the end of this period, the young Officer's commission is, hopefully, confirmed. He can then 'consolidate and develop' his career in the corps of the Royal Marines.

'MARINES IN ACTION'

The new Marine or Officer is posted to one of either of the Commando units, or to a ship-borne detachment (called a Naval Party). Continuation training includes parachuting, and mountain and Arctic warfare skills. The Navy's assault landing craft are also manned by the Royal Marines.

Other tasks for which Marines may be selected and trained include the renowned Special Boat Service (S.B.S.), specialising in 'Cockleshell Heroes' style raids for which the final training standards are above and beyond that of any other military group or element in the world.

The corps is thus a strong, quick-response organisation, which is self-contained and can go anywhere. But make no mistake about it, to be able to 'hack' the job, a Royal Marine Commando must be tough, fit and competent. It is with the aim of producing such men that the selection, initial and continuation training of the corps, is directed. 'Yomping' is one of their words for a forced march of any length over any terrain under any condition. All military units evolve their own slang – which is something they are particularly good at.

In the making and forging of Marines, it is as well to keep in mind that the training methods are continually moulded and kneaded to suit defensive requirements and with World War II massive land battles now less likely and the world now turning to global terror, the training schedules will move and change as the scene dictates, and have indeed changed, since the terrorist events of 11 September 2001 in New York and Washington, but the fundamentals of quality and discipline within the Royal Marines I hope will never change, even if the timings and training pamphlets do.

These training routines which I have run through were standard during the last quarter of the twentieth century, but will by now have changed with time and requirement, but the principles remain the same. Many are called but few are chosen!

If we now fast forward to 2004-5, we find a whole new world of global events never envisaged at the turn of the century.

DEFENCE CUTS 2004

Whilst, however, it is strictly inaccurate to say that the Corps will be unaffected by recent defence cuts, not least because of internal Naval measures to balance their in year costs for the immediate future, all Royal Marines can take considerable satisfaction from the publication 'Delivering Security in a Changing World – Future Capabilities' produced by the Secretary of State for Defence in July 2004. Under the Maritime section of this document, the following text appears:

....As set out in the White Paper, the future Navy will provide a versatile and expeditionary force with an increased emphasis on delivering effect onto land at a time and place of our choosing. The future force structure will be focused on the carrier strike and amphibious capabilities, including the Commando Brigade. In the short term, the capability will be built around the existing carriers and Joint Force Harrier operating the upgraded Harrier GR9. The new carriers deploying the Joint Combat Aircraft (JCA) will transform our capability to project power from the sea. They will have greater reach, sustainability and survivability than the existing carriers and will be able to deploy a much more powerful mix of fast jets and helicopters. The state-of-the-art, multi-role JCA will provide significantly increased performance, improving strike and reconnaissance capabilities, as well as incorporating stealth technology. Similarly, a robust and modern amphibious capability based around two new ships, HMS ALBION and BULWARK, supported by the Bay Class landing ships will provide a step change in our ability to launch and support forces ashore.

....In summary, we are building a versatile, expeditionary maritime capability with far greater and more flexible 'punch' into the land environment, delivered at range from the UK and a time and place of our choosing. This will comprise a transformed strike capability based around future carriers, with Joint Combat Aircraft, and nuclear powered submarines alongside a robust amphibious capability including 3 Commando Brigade. We will also be able to provide appropriate independent forces to smaller scale operations through more individually capable frigates and destroyers, which will also offer networked force protection to the carrier and amphibious task groups. In total, this means a maritime force transformed in its ability to conduct maritime operations and contribute to the land environment in terms of strike, amphibious capability and surveillance.

There is little doubt therefore that in 2004 and for the future, both the Royal Marines and the Royal Navy's amphibious capability are an absolutely key and central part of current and future maritime concepts. In summary, it means that the Corps' future is more secure than the majority of us can remember. The Globe and Laurel has no doubt that this is due to the first class leadership which the Royal Marines continues to enjoy, its integration into the heart of the Fleet organisation and last, but by no means least, the high professional standards achieved by all members of the Corps during operations, deployments and our training since World War 2. The Corps is privileged, we are recognised and the Royal Marines now face the future with certainty.

CHAPTER 1

THE REGIMENT OF THE SEA

"THE GOD OF WAR HATES THOSE WHO HESITATE"

The Commandant of a Royal Marine Training Unit was walking across the barrack square. The Regimental Sergeant-Major had dismissed the morning parade. The recruits were dispersing to their dinner. The Commandant stopped one of them.

"What is the motto of the Corps?" he asked.

Given a little notice of the question the man would, no doubt, have produced the answer his Commandant expected, for the motto which the Corps has proudly carried through the centuries "Per mare per terram" – by sea, by land – is prominently displayed in every Royal Marine establishment. After a moment's hesitation his face cleared and he rose to the occasion, as Marines are trained to do.

"Esprit de Corps, sir," he said.

That answer, I would say, was true in spirit, if not correct in word. It is the clue to the Royal Marine. It shows his attitude to life. He is a soldier who serves on the sea or for the sea, but although he enjoys the fellowship of a ship's company afloat, he never forgets that he is one of (at that time) His Majesty's Royals – "Soldier an' Sailor too".

There is nothing new in this. The Royal Marines' Esprit de Corps has its roots in history. George Forster, the botanist, who sailed with Captain James Cook on his second voyage round the world, has described how, one evening in August 1774, a Marine on board The Resolution fell into the sea while drawing water on the ship's side. He could not swim, but the ship was brought-to and he seized one of the ropes which were flung to him. When he had been hauled on board he was taken below by his comrades, who "gave him a dram or two of brandy to revive the animal spirits, and treated him with all the care and attention necessary to ensure his full and speedy recovery as a result of an 'Esprit de Corps', to which sailors at that time were utter strangers".

Today, this sentiment is something more than pride of regiment alone, for it is the human bond which unites a scattered brotherhood. A Quartermaster, whose father and grandfather had been Marines, and whose son was a Corporal, once said to the Adjutant-General, Royal Marines, "We're not a Corps, sir, we're a family," and this family tradition was once so strong that a naval officer was once moved to observe: "They don't recruit Marines, they breed 'em." Names such as Daly, Owen, Adair, Phillips and Pym, recur again and again in the Corps' records. Some families had three generations serving during the second World War; one had no less than six brothers. There was a non-commissioned officer whose forebears had been serving for 150 years. His great-great-grandfather was a Marine when Nelson was made a Vice-Admiral. The Captain of Marines aboard H.M.S. Victory, Nelson's flagship, was killed at Trafalgar, but one of his descendants was in the Corps in 1944. Daughters as well as sons were eligible to carry on the family tradition, and the Wrens of the Royal Marines (unofficially known as "Marens") wear the Globe and Laurel badge on their caps.

Not all Marines were born in the Corps, but few died out of it, and it made good use of its pensioners' rich experience, gathered afloat and ashore in every corner of the globe, holding that "days should speak, and multitude of years should teach wisdom". On the outbreak of World War Two, these pensioners were called back to the colours up to the age of 55. They were allowed to return to service beyond that age, and many did. One, who was called up as a reservist in the first war, was accepted for service yet again in September 1939. He was then 72. This retired officer, who wore the South African ribbons, then spent a year commanding a battery in Iceland.

These pensioners and retired officers served the Corps well by passing on the torch of tradition to the younger generation, and by instilling into the recruits that habit of obedience that we now call discipline which today regulates our lives and sustains us in adversity. Young Marines learned as much by example as by precept, until they came to take a pride in the Corps and so, as members of it, as much pride in themselves as their predecessors did before them. The Marines have always been meticulous about their appearance. As far back as 1756 a Plymouth Divisional order directed that "No Marine shall appear in the streets without his hat being well cocked, his coat hooked back, his hair combed and tucked and tyed, if long enough, and he is to keep his clothes without dirt, spot, ripp or ragg."

In 1940 the Corps had greater need of its pensioners than ever before. Its strength had been multiplied many times since the outbreak of war. This expansion had compelled it to look far beyond family connections, and the new recruits were drawn from every walk of life. Many came as volunteers. One of them, on being asked if he could swim, looked puzzled but remained undaunted.

"Blimey, haven't you got any ships left!" was all he said.

There was a large proportion of men who joined for hostilities only in the blue-uniformed sea-going detachments. Others, who wore khaki, with the red and blue flash "Royal Marines" on either shoulder, served ashore with the Siege Regiment, who would man the naval guns ashore, with the Mobile Naval Base Defence Organisations, with the Royal Marine Engineers, with the Royal Marine Commandos, or with the numerous other units which had been formed to meet the emergencies of the war.

In spite of those varied commitments, the normal function of the Royal Marines remained constant. He was a soldier who goes by sea and returns by sea. Whether he served afloat or ashore, whether he joined for continuous service or for hostilities only, he was trained for amphibious warfare. It was his boast that he was the first to land and the last to leave.

"Per mare per terram"

Part of his training was to learn something of the story of the Corps he served. When he first entered barracks he would see on the walls of his quarters a big, coloured poster depicting the uniforms of the Royal Marines throughout their history and describing the Corps' badges, distinctions and memorable services. He would learn that the birth of the Corps was on 28th October, 1664, when Charles II sanctioned the formation of a regiment for sea service, to be known as the Duke of York and Albany's Maritime Regiment of Foot, otherwise the Admiral's Regiment, for the Duke was Lord High Admiral of England; that on 24th July 1704 Marines under Sir George Rooke took possession of the Rock of Gibraltar, thereby winning "an immortal honour"; that in 1755 the Corps, having been disbanded after the peace of Aix-le-Chapelle seven years previously, was re-formed and has had a continuous existence ever since. He would find a name familiar to him from present-day communiqués when he read that during the Seven Year War, while the British fleet was blockading Brest, the Marines seized Belle Isle as an advanced base in what would then be called a combined operation, and were granted the privilege of wearing on their colours the Laurel Wreath, which now forms part of their badge.

He would also discover that George III granted the Marines the style of "Royal" on 29th April 1802, "In consideration of their very meritorious service during the late war", and that 25 years later the Duke of Clarence, Lord High Admiral of Great Britain and General of Marines (afterwards William IV), presented the Corps, on behalf of George IV, with new colours displaying the badge which every officer and man wears to this day. His Royal Highness explained that, owing to the difficulty of making a selection from "so many glorious deeds" to inscribe on the colours, their Sovereign had been pleased to adopt "the great Globe itself", encircled with Laurel, as the most appropriate emblem of a Corps whose duties carried them to all parts of the world, "in every quarter of which they had earned Laurels by their valour". His Majesty considered that his own cipher (G.R.IV) should be interlaced with the Foul Anchor, which showed their connection with the Royal Navy. Their motto, "Peculiarly their Own", remained, and surmounting the Imperial Crown on the badge His Majesty directed the word "Gibraltar" to appear, in commemoration of their earliest great distinction.

It was well that the recruit should understand the symbolism of the badge he would wear, for it was compact with history. Since it was presented, the Royal Marines then gained so many fresh honours that, as Admiral Lord Charles Beresford told them in a signal on the 250th anniversary of their foundation: "A thousand colours could not contain their brilliant record."

Some of those achievements, the recruit would learn, commemorated on 16 chosen days of the year, by fanfares sounded on the silver bugles which were presented to the Royal Marine Divisions and the Depot in memory of the officers who fell in the first Great War. From the brief citation read out by the Adjutant on the parade ground, he would learn that Marines fought at the Battle of St. Vincent on St. Valentine's Day 1797, under Sir John Jervis; on the Glorious First of June, 1794, under Admiral Lord Howe, their first Honorary Colonel; at Camperdown, at Nelson's three great victories of the Nile, Copenhagen and Trafalgar, and, coming to the days of the first great war, at the landings in Gallipoli, at the Battle of Jutland, and at Zeebrugge. In those three actions alone the Corps gained four Victoria Crosses.

"Soldier an' Sailor Too"

On ZeebruggeDay 1943, Colonel W. T. Clement, United States Marine Corps, passed out the King's Squad at the Portsmouth Division and listened to the fanfare before he addressed the parade. It was fitting that the chief representative of the United States Corps in Great Britain should take the salute on one of the Royals' most memorable anniversaries, for the two Corps have close associations. The United States Marines, who have a great tradition and a long record of achievement, wear as their badge the other side of the Globe – the Western Hemisphere – and during the Boxer Rebellion, British and American Marines fought under the command of each others' officers. During the Second World War, two United States Marines took the course at Chatham Barracks and at the Royal Marines Military School, and were given direct commissions in their own Corps on passing out.

As Colonel Clement had said, the organisation of the United States Marines is based on that of the Royal Corps, the essence of which is its flexibility.

"There isn't a job on the top o' the earth
The beggar don't know, or do,
You can leave 'im at night on a bald man's 'ead
To paddle 'is own canoe;
'E's a sort of bloomin' cosmopolouse –
Soldier an' sailor too."

So wrote Rudyard Kipling, at a time when the activities of "'Er Majesty's Royals" were less varied than those of the Corps today.

The Royal Marines have fired their guns in nearly every important engagement of the second Great War. Detachments from H. M. ships landed in Norway and Madagascar. Small expeditionary forces, sent from England, occupied the Faeroes and Iceland. At an hour's notice, covering parties went to the Hook of Holland, Boulogne and Calais. Marines formed the rearguard in Crete. Survivors from the Prince of Wales and the Repulse fought ashore in the Malay Peninsula. Volunteers manned river boats in the Burma Campaign. Royal Marine gunners manned some of the Channel batteries and other long-range naval guns ashore, also anti-aircraft batteries on the English coast and in Malta. A special detachment formed a guard for the Prime Minister at the Casablanca Conference. Royal Marine Engineers constructed naval air stations and bases for the fleet. Pensioners, as mentioned before, fought the guns in merchant ships at sea. Royal Marine Commandos were at Dieppe and were the first sea borne troops to land in Sicily. Royal Marine pilots have flown with the Fleet Air Arm in almost every theatre of the last war. And in the anxious summer of 1940, when the strength of the British Army had been shattered, the Royal Marine Brigade waited in readiness to justify Lord St. Vincent's prophetic tribute to the Corps:

"I never knew an appeal to them for honour, courage or loyalty that they did not more than realise my expectations. If ever the real hour of danger should come to England, they will be found the country's sheet anchor."

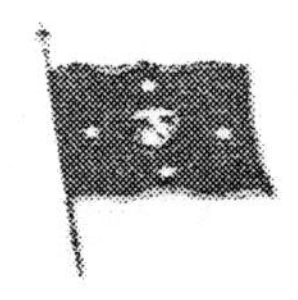

8 September 2003

Tony

Dear General Milton,

It is with great pleasure that I extend birthday greetings to you and your Marines as you celebrate the 339th anniversary of the Royal Marines. Since 1664 the Royal Marines have been your nation's premiere fighting force. From Gibraltar to the Battle of Trafalgar, and most recently in Iraq, your Marines have continued to reinforce the Royal Marine reputation for professionalism and tenacity in combat. During Operation Iraqi Freedom, your Marines performed magnificently and, as always, they serve as a shining example of excellence that is emulated by Marines worldwide.

As you gather to celebrate your storied and illustrious history and pay tribute to those who have paid the ultimate price for freedom, know the United States Marines stand ready to fight by your side when our nations call, in any clime or place.

On behalf of all United States Marines, I wish you and your Marines continued success and hope that you have a joyous and memorable celebration. Per Mare Per Terram and

Semper Fidelis,

Mike

M.W.Hagee
General, U.S. Marine Corps
Commandant of the Marine Corps

Major General A.A. Milton
Commandant General, Royal Marines
Headquarters Royal Marines
West Battery, Whale Island
Portsmouth, Hampshire PO2 8ER
England

best wishes to superb warriors and allies! Mike

CHAPTER 2

BEARERS OF THE SWORD

"IN THE FLEET AND FOR THE FLEET"

The traditional role of the Royal Marines is to reinforce the strength of the Navy at sea. The Sea-Service Marines would man a proportion of the ship's armament. They formed landing parties from ships or shore bases to carry out definite tasks, followed by re-embarkation when their purpose had been achieved. They created and occupied advanced bases on friendly or enemy territory to further the action of the fleet, or in larger bodies to act as striking forces to extend the power and reach of the Navy over the land. Thus the Corps' primary function was to serve in the fleet or for the fleet, and to supply the Admiralty's requirements, which in the Second World War, as in the first, became more insistent as the number of naval vessels increased.

The Divisions at Chatham, Portsmouth, and Plymouth provided the fleet with Royal Marine detachments. Chatham was the Senior Division, which occupied the barracks there since 1780. The Plymouth Marines first marched into Stonehouse Barracks in 1783. The Portsmouth Division had occupied several headquarters, including Fort Cumberland, where specialist and tradesmen were trained, and then moved into the newly-built Eastney Barracks in 1866. The depot at Deal in Kent was opened in 1869, but the building dates back to 1797. The above establishments are all in perfect order today, complete with myriads of subterranean stores, tunnelling, creaking steel doors, and tales of ghostly warriors of times long gone. The history of Fort Cumberland alone would provide a fitting script for the talents of Hammer Horror Films, but that's a story for a dark, wet afternoon and not today!

The Royal Marine barracks and establishments fly the Union Jack, not the White Ensign, and during the last war, although under the control of the Admiralty through the Royal Marine office in London, they are not commissioned as ships like naval shore bases. Marines were paid by the Admiralty, but came under the Army Act Ashore and the Naval Discipline Act Afloat. The dividing line was the ship's gangway.

Each Division had its own character and traditions. Each was a self-supporting community, with its own technicians, from blacksmiths to bricklayers, draughtsmen to divers, miners to camouflage modellers. Other skilled tradesmen belonged to the Royal Marine Engineers, who were organised in companies for the purpose of erecting buildings, constructing harbour works at naval bases and naval air stations at home and abroad. They claimed to be able to put up "anything from a wash hand basin to a pier". The officers were all Civil, Constructional or Mechanical Engineers; the other ranks were craftsmen exercising their own trades. They operated under the Civil Engineer-in-Chief at the Admiralty, and formed a mobile force of experts wherever the fleet had need of them.

In each Division was a tailor's shop, where the blue uniforms for the sea-going Marines were made. Hundreds of thousands of garments were also made for naval ratings, which gave outside employment to many women who were relatives of past or present members of the Corps. Each Division, too, had its printing office, for the Royal Marines were the printers of the Navy and provided the compositors and pressmen who were carried in every flagship. Every one of these specialists, whether ashore or afloat, was also a trained fighting soldier.

The Marines' main duty afloat was with the ship's guns. It had been so since 1804, when, at Lord Nelson's suggestion, the Royal Marines Artillery (with blue tunics) was formed to provide the fleet with trained artillerymen who would, in their turn, train the seamen in naval gunnery. Marines remained the gunnery instructors of the fleet until 1859, when continuous service for seamen was instituted. Those who were retained for cutting out parties ashore and for grappling and boarding parties in battle, were given the title Royal Marine Light Infantry (with red tunics) in 1862, and the Corps was then divided into two branches, the Blue Marines (R.M.A.) and the Red Marines (R.M.L.I.). In 1923, owing to the reduction of the Corps' strength, the two branches merged into the Royal Marines again, and the titles of Gunner and Private were dropped for that of Marine. The blue tunic of the R.M.A. was retained, but the narrow red strip of the Infantry was adopted for the trousers, and the Globe and Laurel emblem was surmounted by the Lion and Crown, instead of the Light Infantry Bugle and the Artillery Grenade.

During the last war, naval gunnery was still the most important part of the sea-going Marine's war training, which also included anti-aircraft gunnery, small arms, searchlights, signalling, Infantry training for rapid landings, and transport work by land and sea.

At each of the Divisions there was a sea-service battery where recruits were taught naval gunnery. At Chatham and Plymouth, the recruits spent their last fortnight in establishments outside the barracks, learning seamanship and naval routing to accustom them to life at sea. They lived in port and starboard watches, slung their hammocks at night, stowed them in the morning, learned the handling of ships boats, and the duties of sentries afloat. Only nautical terms were used and they soon became accustomed to speaking of stairs as "ladders", of half-holidays as "make and mends", of "sleeping onboard" and of "going ashore".

The senior squad at each Division was known as the King's Squad, the best all-round recruit of which was awarded the King's Badge (instituted by King George V in 1918), the Royal Monogram enclosed in a laurel wreath and worn below the left shoulder throughout his service.

"The School of Music"

The only continuous service Marines who were not called upon to serve in H.M. ships were the members of the staff bands at the Divisions. They went to sea only with the Royal Family – the last occasion just before the war being when the Plymouth Band accompanied H.M. King George VI on his visit to Canada in the summer of 1939. In peace-time, the Portsmouth Division provided the band for the Royal Yacht.

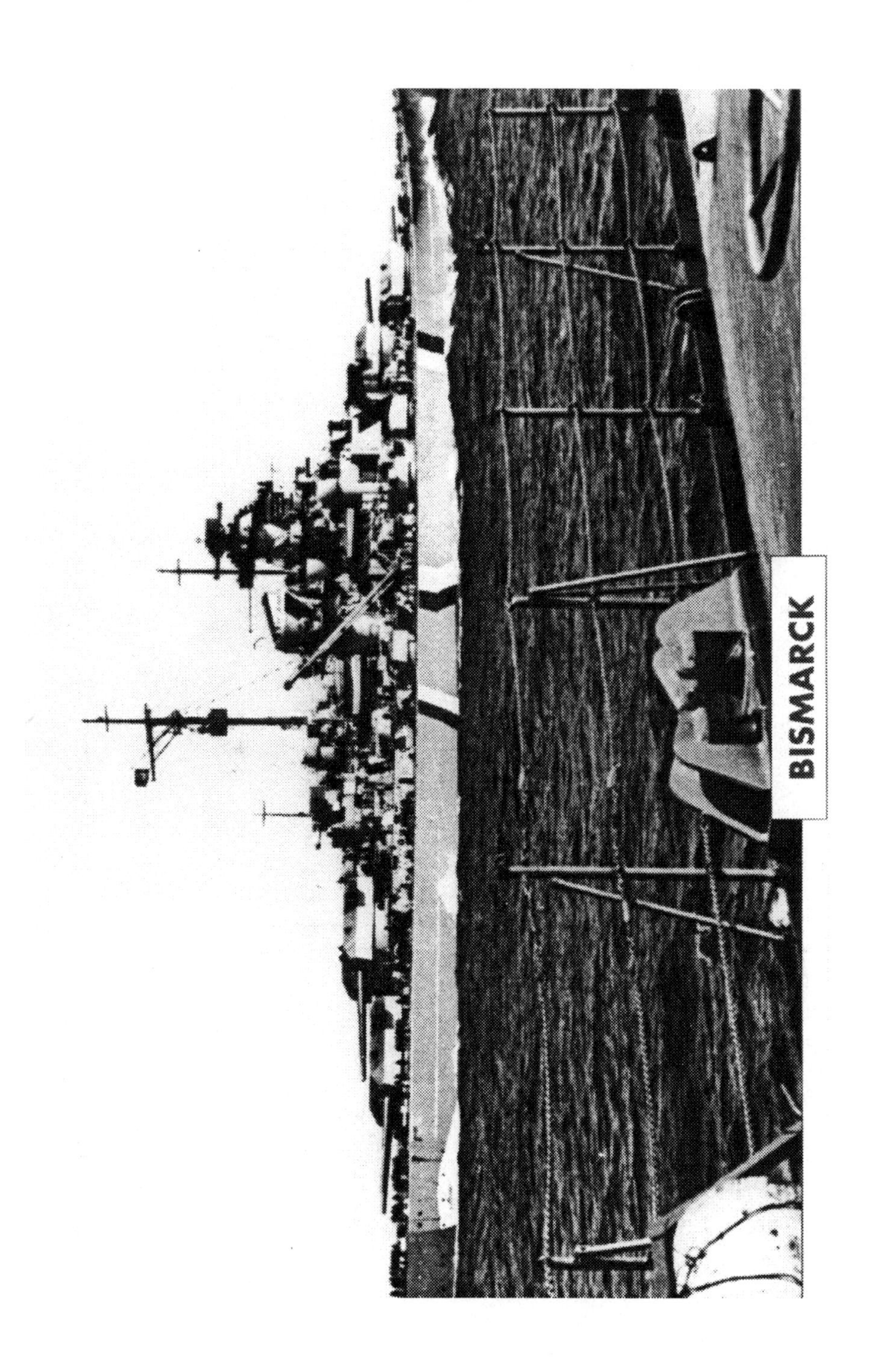
BISMARCK

HMS KING GEORGE V

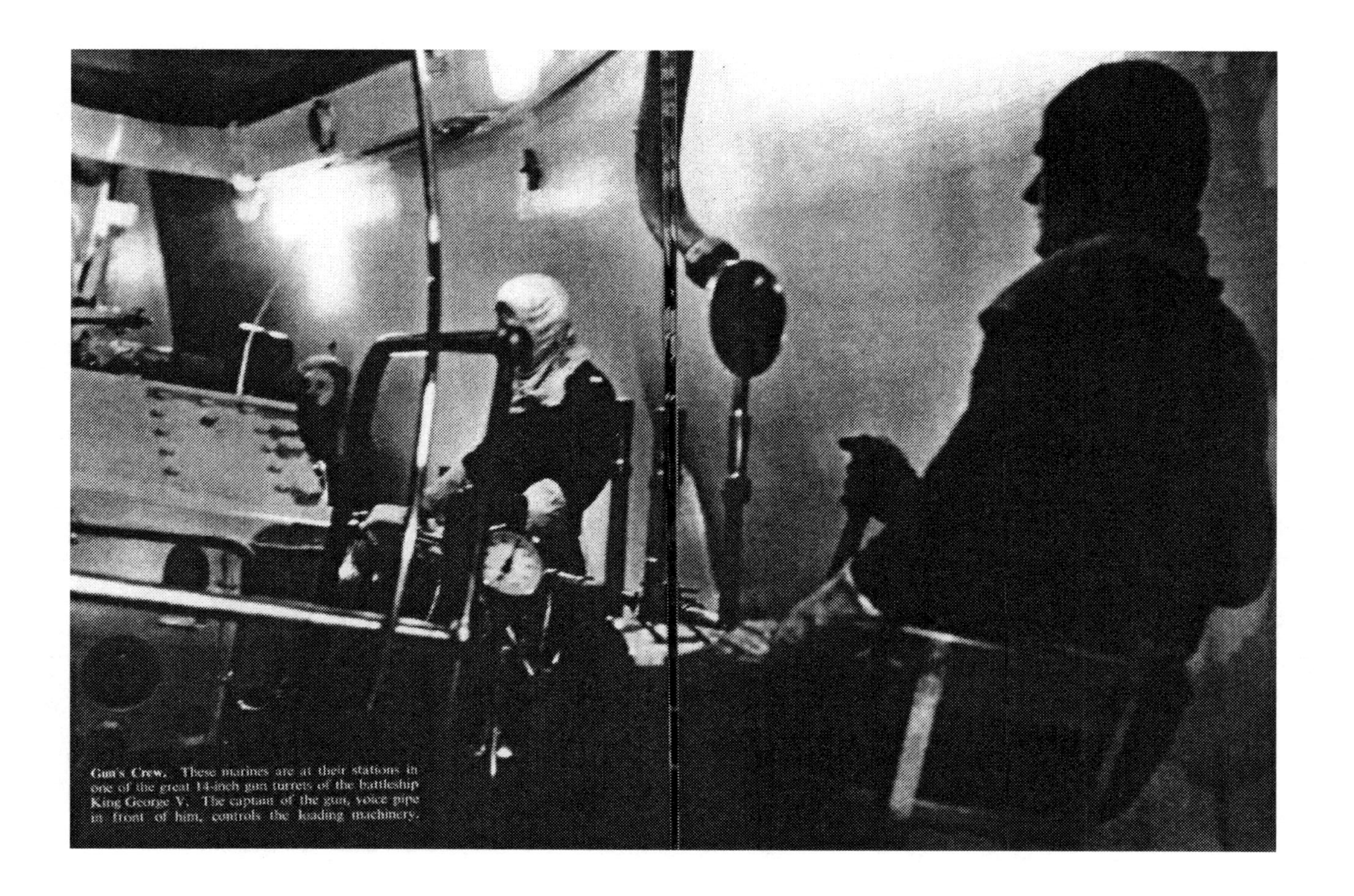

Gun's Crew. These marines are at their stations in one of the great 14-inch gun turrets of the battleship King George V. The captain of the gun, voice pipe in front of him, controls the loading machinery.

Marines at the Guns. Marines man and supply a quarter of the armament of the ships in which they serve. Here two of the Marine detachment in the battleship Howe are working on high-angle 5·25-inch guns. Behind the guns is the crane used for recovering the reconnaissance aircraft carried in the ship.

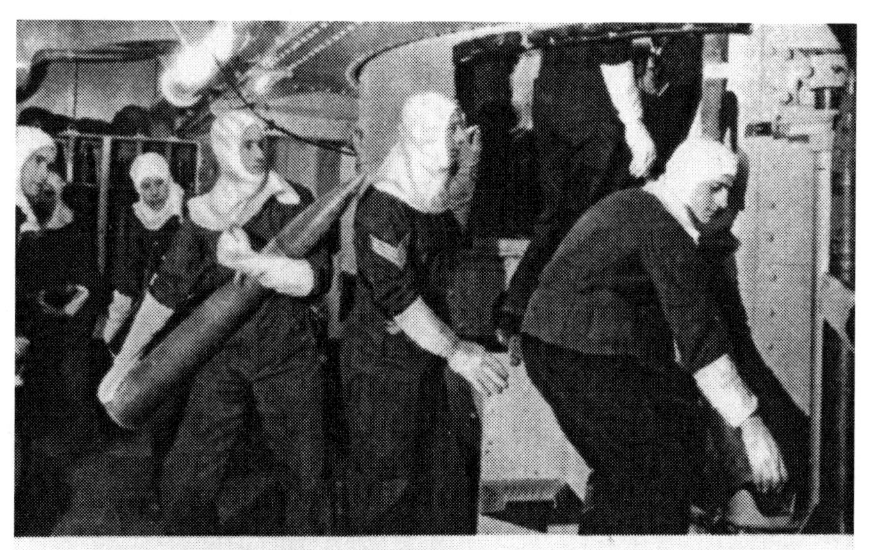

This marine supply party in the aircraft-carrier Illustrious is passing ammunition up to the 4·5-inch guns in the mounting above. The white hoods and gloves, always worn in action, protect them against burns. Where enemy forces may be met, marines live by their guns, as (below) in the gunhouse of a 6-inch turret.

HMS RODNEY

HMS WARSPITE

ADMIRAL GRAF SPEE

These bandsmen were highly trained musicians who had no other duties but air defence. From the earliest days, when the "Band of Music" appeared in white stockings and breeches, with black buckles and garters, the Staff Bands have had permission to play at outside engagements "for their own advantage". They were familiar to the public on parades for Warship Weeks and Wings for Victory campaigns, and have delighted millions of listeners to B.B.C. programmes, not only in Great Britain, but in the Dominions, the United States, and even Germany.

The Drum-Majors of the Staff Bands were responsible for the training and welfare of the Boy-Buglers, who entered the Corps from the age of 14. As well as learning to play the bugle, the drum and the fife, they continued their education under a naval schoolmaster, and had their own block in barracks. There were nearly 100 bugle calls to learn, and some have a different significance ashore and afloat. Their parents' permission was necessary before they could be sent to sea, and many of them performed gallant service during the war. It was a Royal Marine Boy-Bugler who sounded "Action Stations" from the bridge of H.M.S. Exeter in the Battle of the River Plate (his two fellow Buglers were killed when the bridge was put out of action);' another fought ashore in Malaya after the sinking of H.M.S. Prince of Wales.

At sea, the Boy-Buglers were supervised by a fleet Bugle-Corporal, but for discipline came under the Sergeant-Major of the detachment, the strength of which varied from about 180, with a Major or Captain in command, in a battleship, to 40 under a Subaltern in a small cruiser. At the beginning of the Second World War, Royal Marine detachments under Sergeants were carried in the tribal class destroyers to ease the naval manning situation, but towards the end of the war they were confined to battleships, aircraft carriers and cruisers. The Fleet Royal Marine Officer, a Colonel, was a member of the Commander-in-Chief's staff.

MARINES AFLOAT

The officer in command of a detachment afloat had always passed a full Naval Gunnery course, and he was directly responsible to the Captain. His was no easy task. He was often dealing with naval officers far senior to himself. He must have initiative and professional ability, and self-reliance to tackle all manner of problems, for he may be the only Royal Marine officer on board and would have no-one of his own Corps to help him. On him would fall the responsibility of maintaining the Corps' prestige, so that whether it was a matter of commanding his turret in action, leading a landing party and the logistics thereof, storing ship, handling ammunition, painting, drill, parades, or discipline afloat or ashore, he must, and invariably does, see to it that his men uphold their high tradition.

The Marine Detachment forms part of the ship's complement and has its special accommodation between the officers' quarters and the mess desks of the Naval Ratings, an arrangement dating back to the days of sail, when most of the sailors were "pressed" men or "picked up at Hab Nab", as the saying went, and the loyalty of the Marines, the only regular troops on board, could be relied on in case of mutiny. These quarters were known as "barracks". The Marine's reply to an order was "Very good, Sir," instead of the seaman's "Aye, aye." Otherwise, he adopts naval parlance afloat, but does not remove his cap on entering an officer's cabin or on being brought up as a defaulter, as a seaman does, and when the Captain inspects the ship's company at Divisions, the order "Off Caps" is given to the ratings but not to the Marines.

The officer of the Detachment takes charge of the ship's secret and confidential books, and also did ciphering duties. The senior N.C.O. (always known as the Sergeant-Major) was responsible for seeing that the Navigator had wound the ship's chronometers, and reported "Chronometers Wound" to the Captain each and every morning; he also supervises the routine of the Royal Marine Office on board. In all ships which carry Marines, one or two N.C.O.'s, according to the size of the ship, performed the duties of the mail office, with a couple of men to help them. When the ship was in harbour, they would deliver the outgoing mail to the depot ship, collect and sort the incoming mail. In a battleship they could be dealing with 700 items of mail per day, including the official M.O.D. correspondence. "With something like 2,000 men on board, it's a big village," observed the Post Sergeant of one flagship. In the printer's shop of a flagship, there would be a Corporal and four Marines; all the setting was done by hand.

The detachment on board provided the Corporal of the Gangway, who was responsible for seeing that no unauthorised person enters or leaves the ship, and for bringing defaulters before the Officer of the Watch. Marines acted as Admiral's Orderly (in a flagship), and provided the Ship's Guard, which mounted every morning with the band when the colours were hoisted. One of the Guard's duties was to supply a sentry over the keyboard, day and night, a highly responsible task, for there were over 200 keys, including those of the various magazines, and the sentry must see that every key given out is signed for and that no unauthorised person takes one.

Not least among the personnel of the sea-going detachments were the musicians of the Royal Marine Bands, which were distinct from the Divisional Staff Bands and were trained at the Royal Naval School of Music, popularly known as the "School of Wind", founded in 1903. The musicians wore a lyre on the collar of the tunic and the letters R.M.B. on the shoulder-straps.

During the first war, the bands provided by the school were on some 40 ships and their losses were approximately one-tenth of their strength. In the Second World War there were around 100 bands, 70 of which were afloat, the remainder in naval establishments ashore; each could provide a military band, a string orchestra and a dance band, and each had a Bandmaster, first or second class, who ranked as a Chief Petty Officer or Petty Officer. The band came under the Senior Royal Marine Officer of the Detachment for Discipline and Administration, but there was always a Fleet Bandmaster, who ranked as a Warrant Officer. The bands of the Home Fleet had formed a symphony orchestra. On one occasion, within an hour of giving a symphony concert, the musicians were closed up at action stations in their several ships at sea.

In war time, the primary function of a band at sea is to form part of the fighting organisation of the ship. A large proportion would man the transmitting stations, where the gunnery fire-control problems were worked out in the bowels of the ship and transmit to the turrets the gun training and elevation required for hitting the target. Some were in the local fire-control calculators of the turret, other men helped to supply ammunition to the guns. They also acted as stretcher-bearers with the Royal Marine Landing Parties. In the River Plate engagement, when the Exeter's transmitting station was put out of action, the band immediately helped with first aid, or joined up with the fire-fighting parties.

While their ships were at sea, the members of the band usually played no music at all, but performed an arduous round of watch-keeping duties, and on their return to harbour, when their shipmates were able to relax, the musicians would then entertain them. It was remarkable how they retained their enthusiasm under war conditions. There was even one point in time when they played during an air raid on a convoy which was unloading at Malta. One of the merchant ships, with a precious cargo of fuel oil and tinned food, was hit while still laden but, although lying on the bottom, was still above water. To hearten the unloading party – soldiers, and ratings from H.M.S. Penelope - the bandmaster and eleven of the Penelope's musicians went alongside in a lighter craft and played marches and selections from "Snow White", at the same time watching for the hoisting of the red flag which would give fair warning of a raid. Time after time the Stukas "came in like a flight of ducks". Once, as a German bomber dived low over the harbour, the working party faltered for a moment, but the band played on and the unloading continued. As one of the musicians said, "It wasn't much fun trying to get a high note with one eye on a stick of bombs coming towards you," but the Penelope's band played for two days from nine to four, until the cargo was ashore.

"At the Ship's Guns"

Like the musicians, every Marine employed on specialist duties played his part in fighting his ship when "action stations" was sounded, whether he was mailroom, printer, wardroom attendant or orderly, he would assist his comrades in manning and supplying the guns.

A big capital ship usually had four main gun-turrets – A and B, X and Y. The Royal Marines traditionally manned X turret from top to bottom, and one of the secondary armament batteries, port or starboard, as well as a considerable proportion of the anti-aircraft guns. This was the primary duty of Marines afloat. The Royal Marine gunners of the Ajax, Achilles and Exeter helped to batter the Graf Spee when, according to the Sergeant-Major of the Ajax, "Everything went with as much precision as on a peace-time gunnery exercise."

Those of the Warspite played their part in the sinking of the German destroyers at the Second Battle of Narvik, and again at Matapan when the cruiser Zara was sunk ("We gave her a broad-side at short range which lifted the whole top off," as a Colour-Sergeant put it), then the Fiume and the Italian destroyers. Those in the King George V and the Rodney sent their shells into the Bismark. The Rodney detachment alone fired 132 shells, after waiting in apprehension for six salvoes before X turret would bear. "All the guns' crews were keyed up, hearing the other turrets firing and afraid they might miss the party," the Gunnery Instructor said afterwards. Throughout the action, the Officer of the Turret was seated at his periscope, passing back a running commentary to the Captains of the Guns and their crews, through the din and clatter of the loading machinery. "She's on fire for'ard" – "Her after turret has been hit by our shells" – "Her people are leaving the ship" – "Her deck is black with them." By the end of the engagement, the guns' crews had been in the turret for twelve hours.

Nor had the Royal Marines Gunners been targeting enemy warships only, for they helped to destroy innumerable German and Italian merchantmen, and took part in the bombardment of land targets from Norway to North Africa.

The shallow-draught monitor, H.M.S. Terror (Commander H. J. Hayes, D.S.O., D.S.C., R.N.) which took part in many such bombardments, had her one 15-inch twin turret entirely manned by Royal Marines, except for some supply numbers down below, and a Royal Marine officer, Captain R.R.G. Hoare, was Gunnery Officer of the ship. From June to November 1940, the Terror was moored in Lazaretto Creek, Malta, as part of the defences of the island, and during the first month alone she engaged enemy aircraft formations on more than 50 occasions. No 15-inch shells were fired, but the Royal Marine Gunners took their turn with the naval ratings in the 4-inch anti-aircraft batteries. The Terror performed the same function during the opening of the naval base at Suda Bay in Crete, then took part in General Wavell's offensive in the Western Desert, bombarding Halfaya and Sollum Passes with her main armament from dawn to dusk and denying them to the enemy during a vital period. Later she attacked other targets on the coast, including Bardia, her last bombardment being on the Western defences of Tobruk by night, in

January 1941. By that time both her guns were very worn – they had fired more rounds than any other 15-inch turret in the Navy – and she was then employed on coast and anti-aircraft defence of harbours until she was sunk in a dive-bombing attack on 25th February.

Royal Marine Gunners shot down countless Axis aircraft, not only during the defence of Malta and the invasion of Crete, but while escorting great convoys to Russia and to Malta, and manning the armament of carriers. During the passage of one Malta convoy, one of H.M.S. Indomitable's turrets was wrecked by a bomb, and had to be evacuated by the Royal Marines. Two wounded men remained behind. In spite of the danger of exploding ammunition, Marine G. W. Wright returned to the blazing turret and got one of his comrades out. Although he had been badly burned, he insisted on going back for the second casualty. He removed the man's burning overalls, wrapping him in his own, but the Marine was mortally wounded; he told Wright to save himself, and finally the advancing flames forced him to withdraw. For this loyalty and devotion to his comrades, Marine Wright was awarded the D.S.M.

The Laurels for these actions the Royal Marines share with the Seaman Gunners of the vessels in which they serve; and in the great ships which were lost they also shared the same perils and paid the same price as the naval ratings alongside whom they fought their guns until the end.

"Slinging His Own Hammock"

Their function as Infantry Afloat, however, was theirs alone, and it was always a great moment when they heard the piping of the call "Royal Marines landing party, fall in on the quarterdeck." The Marines manned so important a part of the ship's armament that the number available for emergency landings were restricted by the requirements for manning the guns. But within ten minutes a party would be in readiness with equipment, ammunition and food, the men "hanging on a split yarn" in their eagerness to disembark.

There have been occasions, too, when part of a Royal Marine detachment was employed as a prize crew on board a captured enemy vessel. In February 1941, after the aircraft of H.M.S. Hermes had stopped the 8,000-ton Italian merchant ship Leonardo da Vinci off the east coast of Africa, a party of Marines, with some seamen and engine-room artificers were put on board, under an R.N.V.R. Lieutenant. The Marines rounded up the ship's company – over 80 men – marshalled them in the saloon, and put sentries on the door. Fifty Germans who had scuttled their ship and had been taken off by H.M.S. Hawkins were then put on board the Italian prize. Since the Germans protested against being with the Italians, they were packed into the sick bay, the Captain and the officers being given cabins. During the passage to Mombasa, the Marines helped to work the tow and made their own cooking arrangements. One killed a bullock found on board, another skinned and prepared it. Sergeant C. F. Evans gave first-aid treatment to the Germans, whose hands had been burned by the ropes as they went overboard. He also kept a watch on the bridge to enable the Officer in Charge to obtain some rest.

This incident is a fair illustration of the old naval saying that a Marine must "learn to sling his own hammock", which means that he must learn to look after himself and turn his hand to any job he may be called upon to do. In the art of improvisation he is perfectly instructed, and his versatility makes the Royal Marine detachment a vital part of a ship's company, whether she be in harbour or at sea.

CHAPTER 3

THE ENEMY FORESTALLED: THE FAEROES AND ICELAND

"SHADOWS IN A LANDSCAPE"

The Royal Marines' training for amphibious operations made those who were ashore awaiting drafting to a sea-going detachment, appropriate troops for an occupation force at short notice. This must have been the reason why, after the German invasion of Denmark and Norway had begun and the position of the Faeroe Islands and Iceland appeared hazardous, units of the Corps were chosen to garrison the islands against enemy landings. It had to be seen as a matter of strategic necessity. The occupation of the Danish dependency of the Faeroes was the more urgent, for those twenty rocky, cloud-shrouded islands, rising sheer from the sea, lay no more than 200 miles from the Shetlands.

At 11 pm on 10th April 1940, the day after the Germans had marched into Denmark, Lieutenant-Colonel T.B.W. Sandall was informed that he had been appointed to command a force of Royal Marines, which was to proceed to the Faeroes immediately. Well, I think at this point in our story we have established that Marines are trained to move quickly. By 6 pm on the 12th, Force Sandall, consisting of 13 officers and 180 N.C.O.'s and men from the Chatham, Portsmouth and Plymouth Divisions, had embarked in H.M.S. Suffolk, carrying stores for one month. Steaming at high speed, the Suffolk made the passage in 21 hours and reached Thorshaven, the capital of the Faeroes, at 3 pm next day. Colonel Sandall went ashore with a small staff (but no armed guard), met the British Consul, and made arrangements for billeting. Assisted by a couple of British trawlers, requisitioned on the spot, the whole Force had landed by 8 pm. One company was billeted in a theatre, another in a dance hall. Headquarters occupied a club over the local cinema. There was no opposition. Oh, the luxury of it all as my mind drifts back to my first "foray" onto the streets of Northern Ireland in early 1970. My first "billet" was a burnt-out Ulster bus with not much more than a ground-sheet to keep the weather out, but this only lasted for 12 hours before I was evicted by some very irate locals who berated me as a British Imperialistic raider with the morals of the Hun, so I was forced out of my bus and onto the floor in the corner of a flax mill. "A day in the life of a soldier" it seldom changes; I must tell you more about this sometime. But for now, let's not digress so to get back to the Faeroes – the townspeople gathered and watched the sentries on their beat, much as a London crowd watched the guard outside Buckingham Palace.

Meanwhile, Colonel Sandall had called on the Governor and informed him that the Force had come to ensure the safety of the population and the sovereignty of the islands against possible German aggression. The Governor declared that he must make a formal protest against the occupation. Colonel Sandall undertook to forward this in writing to the British Consul for transmission to the Foreign Office. Until he received further orders, however, his troops would remain in the islands and carry out such defensive measures as he considered necessary.

This conversation was carried on through an interpreter, the Governor leaving the room several times to consult his advisors. When the interview came to an end and Colonel Sandall was preparing to withdraw, the Governor invited him, in excellent English, to join him in a whisky and soda. Diplomatic conventions had been upheld, and from that moment the friendliest relations existed, not only between His Excellency and the Commanding Officer of Force Sandall, but between the local inhabitants and the troops. The Governor had quickly realised that, since he was powerless to prevent the occupation, it was his duty to do all he could to ease the situation for his people, who responded loyally to his appeal. He issued a proclamation declaring the occupation was exclusively for military reasons and would not change the political status of the islands or interfere with the municipal administration, and called upon the population to put no hindrance in the way of the troops. Although censorship was contrary to the Danish constitution, its necessity was accepted with a good grace.

On the following day, the Royal Marines occupied the wireless station and the harbour. Two parties took up defensive positions in the country, which was as bare and desolate as the Orkneys, as mountainous as Norway. The Marines soon adapted themselves to their new conditions. They admired the neatness of the towns, the cleanliness of the houses – painted white, with red roofs – and the churches with their slate-blue steeples. The local schoolmaster acted as interpreter and the Marines played two football matches against the inhabitants – and lost them both. The villagers gave them high tea afterwards, and entertained them with dances and national songs. When the Force was relieved by a battalion of the Lovat Scouts on 27th May, Colonel Sandall was able to report that the behaviour and discipline of the Marines had been excellent and that there had not been a single "incident" during the period of occupation.

Although the occupation of the Faeroes was uneventful, as a perfectly planned and faultlessly executed operation often is, Force Sandall had performed a service which was as valuable as if it had won a battle. In fact, it is not too much to say that it had won a battle, thanks to the flexibility of the Corps, its ability to move at short notice, and its discipline. Looking back now, the Germans had been forestalled. Had there been delay, there is little doubt that the enemy would have invaded the islands, and the effect of a hostile naval and seaplane base so close to the north of Scotland would have been disastrous.

"The Marines Land in Iceland"

The position of Iceland, which is larger than Ireland and commands the North Atlantic trade route, was even more important. Its occupation as a preventative to German invasion took place while Force Sandall was in the Faeroes. The occupying Force was composed of Royal Marines, not from sea-service detachments but from the newly-formed Royal Marine Brigade, the forerunner for 3 Commando Brigade which we read about today and refer to as (3 CDO. BDE. R.M.). In its very early days it consisted of one Infantry battalion, a battery of 4-inch guns and one of four 2-pounders, in all some 30 officers and 650 other ranks, with one naval Howitzer battery. The force was commanded by Colonel (later Major-General) R. G. Sturges and sailed in the cruisers Berwick and Glasgow, with the destroyers Fearless and Fortune in company, on 8th May. Mr. C. Howard Smith, C.M.G., Minister-Designate to Iceland, who had left Denmark when the Germans marched in, accompanied the Force.

At 2.30 am on the 10th, when the squadron was within 30 miles of Reykjavik, the Berwick's Walrus aircraft was flown off to reconnoitre the harbour, and the Fortune made an anti-submarine sweep of the approaches. The cruisers anchored half a mile from the harbour and the Fearless took ashore the first flight of troops, including the Royal Marine detachment from H.M.S. Berwick. A violent snowstorm came on, so that the Berwick was able to steam into the inner harbour and go alongside the main jetty in the centre of the town almost before she had been seen from the shore. The berth was clear and there was no opposition. As one of the officers observed, the landing "was like getting out of a train on to a platform". This was shortly after 4 am. It was by then quite light.

The landing party was met by a few civilians, mainly British residents who spoke the language. They acted as guides, and the Marines lost no time in posting pickets on the three main roads leading from the town, occupying the post and telegraph office, the main telephone exchange, and the broadcasting station. Cars were commandeered, and all objectives were secured without interference from the local inhabitants or the police.

"A Visit to Herr Gerlach"

One of the first objectives was the occupation of the German Consulate. Two platoons, under Major S. G. Cutler, had been detailed for this purpose. They reached the Consulate expecting resistance. The Consul-General, Herr Gerlach, was known to be an ardent member of the Nazi party, and some time before the war the Swiss government had requested his removal from Berne. In response to Major Cutler's knocking, he opened the door himself. He was fully dressed, even at that hour of the morning. On seeing the Marines he protested, but led the way into the hall. One important object of occupying the Consulate was to impound the secret documents. To prevent them from being burned, Major Cutler had thoughtfully brought with him a phomene fire-extinguisher. Having dealt with the Consul, he was about to search the cellar, where he thought the papers might be stowed, when a Marine called out, "Fire on the first floor!" He dashed upstairs, to find flames rising to a height of 20 feet. The Consul's wife and elder daughter was rushing about in their nightgowns, throwing all the confidential books and secret documents into the bath, which had five inches of paraffin in it and was blazing. Well, reading this today in the 21st century you could be forgiven for thinking this to be an ideal situation for some hapless Lance-Corporal to run about waving his hands around and shouting, "Don't panic, Major Cutler, don't panic," but no, the Marines ripped the clothes off the bed in the Consul's bedroom, flung them on the bath and brought the phomene into action. They extinguished the fire in a few seconds, thereby saving most of the papers.

Herr Gerlach was then escorted round the house while a search was made for possible booby-traps by opening drawers, taking up carpets, and moving furniture. The other members of the Consulate were given time to pack two suitcases each, but were warned that if they attempted to secrete any books or papers, they would be allowed to take nothing.

They were then assembled in the hall. Herr Gerlach asked to be allowed to fetch his overcoat from the cloakroom. He was escorted by Major Cutler and a Sergeant. As he reached for the coat, his left hand went to the pocket. Major Cutler seized his arm and took a loaded revolver from his overcoat. Herr Gerlach was then placed under an armed guard.

Cars were commandeered and by 8.45 the Consul, his family and his staff were embarked on board H.M.S. Glasgow. The ensign of the German Consulate was hauled down and was subsequently signed by the senior N.C.O.'s of the battalion. It now hangs in the Royal Marines museum at Eastney Barracks in Hampshire.

There were, however, other Nazis in Iceland besides the Consul-General. Colonel Sturges had a black list of some 70 agents. Within 45 minutes of the Force's landing, the majority had been arrested and were sent back to the cruisers. They included a number of Germans who had come from a so-called ship-wrecked merchant vessel; an officer from a U-boat who had been landed "to look after them" was arrested later and returned to England with Force Sturges.

"Force Sturges Hands Over"

By 09.30 on the morning of the landing, Colonel Sturges was able to report "All quiet, inhabitant friendly". Trawlers were requisitioned to disembark the stores. One of these was the Faraday, of Hull. Less than 15 minutes after the Marines boarded her, she was steaming towards the Berwick for her first load. Her crew joined in the work of unloading and her skipper refused to accept the requisitioning chit which entitled her, him and his crew to compensation. The skipper and his men were glad to work for nothing. All stores and equipment were ashore by 5.30. The presence of H.M. ships being no longer necessary, the squadron weighed an hour later.

As precautions against a German airborne invasion, units of the Force took over the seaplane station, the local glider club and a possible landing ground 43 miles outside Reykjavik. The 2-pounder anti-aircraft pom-poms were mounted on the outskirts of the town. This bloodless blitzkrieg was accomplished, as an officer put it, without a single shot having been fired – even by accident.

The Icelanders accepted the situation philosophically. They, like the Faeroes, were and are an independent people and it was not to be supposed that they welcomed the occupation of their island. There was one party which was strongly pro-British, another equally pro-German – for the Germans had carefully cultivated relations with Iceland in the years before the war. The attitude of the remainder was "Iceland for the Icelanders". Nevertheless, they recognised that the Germans might have come if the British had not, and they regarded the British as the lesser of the two evils. The tact with which the delicate situation was handled, however, the declaration that the island's political status would not be affected, and the excellent behaviour of their self-invited guests, led to a breaking down of reserve. As in the Faeroes, the inhabitants were soon on friendly terms with the Marines, and before the last of the Force had left, more than one British-Icelandic marriage had been arranged.

Had the main Force stayed longer, these relations would doubtless have grown more friendly still. But, with the occupation complete and the situation well in hand, the work of the Royal Marines was done. Like Force Sandall in the Faeroes, by reaching Iceland in time they had kept the Germans out and had made possible the operational base which was to become so important to the Atlantic convoys. The Infantry were spared the hardships, the long darkness of winter and the icy winds, which those who followed them were to know. On 17th May two Cunarders, the Franconia and the Lancastria, arrived with an Infantry Brigade to relieve them.

The expedition had been a remarkable achievement. Six days after receiving the warning order in England, the Infantry of the Force was deployed in Iceland over a front of 70 miles, with a detachment on its way to Akuruyri, 200 miles distant. The batteries were in position soon afterwards. On 20th May, 11 days after the landing, the command had been transferred to the Army, and the Infantry of Force Sturges, re-embarked in the transports, was steaming back towards the United Kingdom.

After the occupation, the Germans broadcast a statement that there would not be a single British soldier left in Iceland in ten days time. Colonel Sturges's answer was included in his Defence Orders:

"There is only one scale of resistance – to the last round and the last bayonet."

The threat did not materialise, however, and our bases and those of our allies remained intact. The initial credit must go to the Corps for its despatch in handling the difficult situation, and it may well be proud to include Iceland as another battle honour which its colours cannot contain.

CHAPTER 4

NORWAY: FIRST TO LAND AND LAST TO LEAVE

"KEEPING THEIR FEARS TO THEMSELVES BUT SHARING THEIR COURAGE WITH OTHERS"

When Norway fell, the people of that small nation knew there would always be another chance to rise again with a little help from her friends. Always another chance, because this temporary wound we call "failure" was not the falling down, but the staying down. Staying down was not part of their itinerary. The Norwegian campaign, which ebbed and flowed during the occupation of the Faeroes and Iceland, gave the Royal Marines opportunities of fulfilling many of their functions. Although it is impossible to list here a detailed account of all their activities, this may be said: True to their tradition, they were the first British troops to land and the last to leave.

The first Marines ashore in Norway after the German invasion had begun were detachments from the cruisers Sheffield and Glasgow. They landed at Namsos (see map) on 16th April 1940, seven days after the Germans had entered the country. They secured the harbour and the road bridges to ensure an unopposed landing for the larger military force, which was to follow two days later. These detachments were under the command of Captain W. F. Edds, the Senior Royal Marine Officer of the 18th Cruiser Squadron, and although the men had been cooped up in ships for several months, they marched long distances through snow-covered roads, then worked feverishly through the night preparing their positions, sleeping in log huts which gave scant protection from the piercing winds, and during the day remained cramped in these shelters to avoid observation from the air. Having covered the landing of the expeditionary force under Major-General A. Carton de Wiart, V.C., the detachment, their task completed, returned to their ships to some welcome relief from Arctic field conditions.

At this point, I must diversify again to remind you that in 1940 the "Mountain and Arctic Warfare Cadre" Royal Marines, which emerged in the later part of the century, had not been formed or even thought of as a permanent training force as yet. It is worth remembering the field operations carried out in Arctic regions were done so with woefully inappropriate equipment and clothing, and the problems of maintaining men in the field for militarily worthwhile periods were fraught with logistical, if not medical, problems. It was not until late '69 while stationed at Stonehouse Barracks, Plymouth on my return from J.W.S. (Jungle Warfare School) in Malaya when 40 Commando were in Singapore, that I was informed that I (now in 45 Commando) would be moving to Arbroath on the magnificently wild and windy east coast of Scotland, a country and a people who will always remain dear to me and who have always had a unique working relationship with the Royal Marines. During this move was born the "Mountain and Arctic Warfare Cadre", a soulless band of cold-blooded warriors formed to train the rest of the unit to the highest standards of M. & A.W. At the head of the training staff was the M. & A.W. Cadre O.C. (Officer Commanding) Captain Mike McMullan, a veritable Grizzly Adams of a man who had, over a number of years, become adept to living life to its outer margins of sanity. But Capt. Mike McMullan and freezing nights of minus 20 degrees went together in Norway as appropriately as Heinz and the tomato, which made him the ideal man to produce warriors ideally suited to protecting the northern flanks of N.A.T.O., as it was to become known. I am convinced even today just as I was then, that his training methods were of such a nature that if the good, quiet people of Arbroath thought we were rather odd, then the Norwegians were soon convinced we were quite irretrievably mad.

In those early days, the one positive advance to M. & A.W. training was the huge amount of money which was made available for the purchase of appropriate equipment for the task at hand.

In 1969, as it happened, a "Major Rod Tuck", who was a very proficient cross-country skier and winter enthusiast, was about to leave the Royal Marines and was somehow commandeered, or if you like "talked" into taking on the tactically complicated job of liasing with other countries who suffered much harsher winters than Britain, such as Canada, Sweden and Norway and select and test, in a series of field trials, a whole range of equipment from clothing to stores which would enable troops to not only live and operate, but more importantly "survive" in the field in the harshest of conditions. Needless to say, these trials took a number of winters in Norway and it was 1973 before, in my opinion, we had a range of equipment which when used properly in conjunction with professional M. & A.W. training methods could maintain troops in the field under operational conditions for any credible length of time.

I have trained in Desert, Jungle, and Mountain and Arctic Warfare with the best equipment that money can buy for any of these theatres, and if you gave me a choice into which you would force me for a month into any one of these environments, with the up-to-date range of Royal Marine M. & A.W. clothing and stores, I would choose the far-flung Northern Tundras of Norway in January. I would certainly prefer it to a U.K. soldier's green uniform on Dartmoor in January where a zero or plus one temperature can penetrate an insidious dampness through a man in a very short period, and render him a liability to his section or troop. But with the proper equipment and observing the tried and trusted methods of M. & A.W. training, I would choose this because I could turn a dry minus 20 degrees completely to my advantage. These skills take time, money, equipment, training, perseverance, and monumental patience with the planning of stores and logistics; for to be caught short with the downfall of one or more members can completely destroy the objective of a long-range patrol through the moonscape of an Arctic night.

You must forgive me for digression, but in trying to explain historic events, it is sometimes worthwhile drawing on comparable situations to bring home a poInt. From the first class stores and equipment of the present day, let us now forget digression and return to those ill-equipped Marines of 1940.

"Holding on at Aandalsnes"

Another force, some 700 strong, with an anti-aircraft battery, was given notice on 13th April to embark for Norway, with the object of seizing the port and railhead of Aandalsnes and preventing the landing of German troops by seaplane, submarine and parachute. It was known as Force Primrose and was composed mainly of detachments from the Nelson, Hood and Barham, which were then refitting. Lieutenant-Colonel (later Major-General) H. W. Simpson was in command. Travelling independently from Portsmouth and Plymouth, the Force had sailed from Rosyth in four sloops by 4 a.m. on 15th April.

The flotilla, steaming with all possible speed, arrived at the mouth of Romsdals Fiord, which leads to Aandalsnes, on the 17th. Then the four sloops went forward cautiously in line ahead, led by the Black Swan (Captain A. L. Poland, D..S.C., R.N.), which carried the Hood detachment. There was no knowing whether the town was occupied by the enemy or not. The Marines stood-to in silence on the upper decks. The fiord was very narrow, and they could see the mountains rising almost sheer out of the water on either side. The night was fine, but intensely cold.

It was almost pitch dark when the sloops anchored off Aandalsnes. The actual landing was an anti-climax after the grim preparation for battle. Instead of German opposition, they found a small crowd of friendly Norwegians waiting to welcome them. There was a strong quay, with a travelling crane to help unload stores and equipment. It was said that there were no Germans in the neighbourhood; nevertheless, Colonel Simpson took the precaution of arranging defensive posts near the jetty. About 2 a.m. the Marines moved off in lorries to take up their positions. One detachment went to a military camp, Setnesmoen, where they spent the night; another was sent to Aalesund, under the command of Major H. Lumley, with a battery of coast artillery, to prevent enemy traffic passing down between the islands and the mainland. Halfway between Trondheim in the north and Bergen in the south, the Central Norwegian Plain is joined to the waters of Romsdals Fiord by a narrow valley, more than 60 miles in length and flanked by precipitous hills. As the valley approaches the shores of the fiord, it opens out into the shape of a fan. In one corner of the fan, fronting the fiord, is the little town of Aandalsnes, in the other corner the village of Veblungsnes. The camp of Setnesmoen is – or was – situated at the point where the valley starts to widen – a collection of long white wooden huts, with low roofs and verandas. When the Marines reached it, the camp was in thick snow.

The task of Force Primrose was similar to that allotted to the Sheffield and Glasgow detachments at Namsos, but on a larger scale: To occupy Aandalsnes and district and to make arrangements for its use as a port of disembarkation for the Central Norwegian Expeditionary Force, the advanced units of which were expected that day. Seamen platoons were to provide working parties for unloading the ships. The Royal Marine detachments occupied a perimeter beyond the town. One of the Hood platoons covered the village of Veblungsnes.

Veblungsnes was one of those typical Norwegian villages which became familiar to the British troops in that distressful campaign. On two sides it was washed by the quiet waters of the fiord; behind it rose a grassy slope which soon merged into the steep sides of a snow-covered and sparsely-wooded mountain. A small wooden church stood back from the Aandalsnes road in a tree-shaded cemetery which sloped down to the fiord's edge. Beyond the church, the road was fringed with houses, some of wood, some of stone, low and red-roofed, with narrow doorways and thick framed windows, a few with fenced front gardens. Off the village square was a solidly built stone house with steps up to the door, a tall tree shading the entrance. It was a tailoring factory. Inside were piles of cloth, ready-made suits and sewing machines. Here the Hood platoon took up their headquarters.

"Field Days for the Luftwaffe"

The main Expeditionary Force began to disembark on 18th April and moved forward to Dombaas, leaving Force Primrose to hold Aandalsnes and its vital landing place and railroad. It was not long before the Germans discovered the occupation, and the Royal Marines' positions were bombed every day.

There were no airfields in British hands from which fighter cover could be provided, and the Germans had complete mastery of the air throughout the operation. But the Royal Marines anti-aircraft guns met them with spirited fire and the Black Swan gave protection to shipping from the fiord. One day, she was bombed for three hours continuously. The Hood platoon at Veblungsnes watched her from their trenches as she dodged and twisted in the narrow waters of the fiord, at times almost hidden by splashes from near-misses, but keeping up an incessant barrage. When at last the raiders departed, without having scored a single hit, the little ship steamed triumphantly past Veblungsnes on her way to re-ammunition at Aandalsnes. The Hood platoon jumped from their trenches and cheered her. Their voices must have carried across the water, for she flashed an Aldis lamp in answer as she passed.

The raids would stop for the day at six o'clock in the evening and three hours of peace would follow. As the sky reddened with sunset and the mountain tops turned from white to gold, a party of Norwegians would gather on the square of grass in front of the Royal Marines quarters in Veblungsnes and sing English songs in their own language to a couple of guitars, which they passed from hand to hand.

"It is almost impossible to put into words," wrote one of the officers, "the feeling of peace that descends on Veblungsnes on these indescribably lovely Norwegian evenings, when one can relax, and smoke, and watch the shadows of the little white houses lengthen, and the sky turn from blue to crimson, without listening to the faraway drone of Heinkels. The loveliness is made all the more poignant by the realisation of horror and death only a short way up the great valley, and by the knowledge that tomorrow the bombing will start all over again, and that this may be the last evening we shall ever see."

It was not long before it became clear that the Central Norwegian campaign had failed. The deciding factor was the German domination of the air. On 29th April came orders for the evacuation. Cruisers and destroyers were coming up the fiord that night and the following day to embark the troops. Force Primrose was to form the rearguard and to hold the mouth of the valley while the evacuation was in progress.

By that time there were but two or three scattered houses still standing in Aandalsnes. Setnesmoen Camp was destroyed on the 29th. In the early hours of the 30th the Marines took up their rearguard positions, which they had orders to hold for the next 24 hours.

“Those hours were the longest I have ever known,” wrote the same officer. “When dawn came we lay up on our hillside while the Heinkels and Dorniers had a positive field-day over Veblungsnes and Aandalsnes. We then could see them through the tree-tops only a few hundred feet up, firing burst after burst of machine-gun and cannon fire into Setnesmoen.”

“Towards midday we had a feast of chocolate and cold baked beans, washed down with melted snow. When darkness fell again, it was bitterly cold. Greatcoats gave hardly any protection and were inappropriate and cumbersome when wet. Head-dress and boots were useless against the penetrating cold and sleep under such conditions was not on the cards as constant trench fidgeting of fingers in wet gloves, and toes in damp socks was the only positive way of obtaining that all-important blood-flow to avoid frost-bite, and we dared not take exercise for fear of giving our positions away to the air or the German scouts in the surrounding woods... the time dragged on. Every minute it seemed to get colder and colder, and the hands of my watch to go slower and slower. From the road along the other side of the valley we could hear faintly the sound of lorries bringing back the remnants of the Army; a comforting sound in as much as it indicated the proximity of friends, even though they might be tired, beaten, and in full retreat.”

“As the darkness deepened, the air raids became fewer and fewer and finally stopped altogether. The valley grew silent and still. Little night noises made themselves heard; the stirring of the tree-tops in the faint breeze; the rustle of some animal in the undergrowth; the far-away hum of a motor-cycle bringing a last lonely despatch-rider back from the front; my watch ticking away on my wrist; the slow measured breathing of the men hidden in the darkness and the endless fidgeting of hands and numb feet in freezing thin leather boots. What was the time? Half an hour after midnight. Another two and a half hours to go. Were we going to get away with it altogether? At last two black shapes loomed panting out of the gloom. Orders. ‘Withdraw at once. Rendezvous Company Headquarters. Hurry’.”

The men rose, stretching cramped limbs. They formed up in sections, then slipped and tumbled down the slope they had climbed twenty hours before. When they reached the rendezvous they were told that the destroyers were waiting, and this information alone sent morale soaring, for no matter how brief a period, it meant hot food, hot tea, hot water, a thawing period for chilled bones, and then on to a briefing for the next challenge. The ships had been ordered to sail not later than 1 a.m. Hearing that there were Marines to be withdrawn, they had remained beyond their time.

The Marines moved off, a long line of men, black shadows on the snow in the moonlight. They came to the beach. Two destroyers were lying 200 yards off shore. The scene was flung into harsh relief by the glare of their searchlights. The men were dazzled and blinded as they emerged from the dim Norwegian night. They moved knee-deep in the icy water and scrambled into the waiting boats. The destroyers gathered way as they drew alongside.

Among the last of the British forces to leave Central Norway was a Royal Marine Howitzer Battery, which had been operating in the Namsos area. The Commanding Officer received orders to withdraw immediately, leaving his guns and all equipment. He replied that he would evacuate with all speed, bringing his guns with him. A further signal told him that the destroyer could not wait. He answered that nevertheless he and his men would remain with the Battery until it was ready to be moved. Although he kept the destroyer waiting twenty minutes, by requisitioning every boat he could find he succeeded in bringing away every gun, all his equipment, and every man.

"It is not the policy of the Corps to leave its equipment in enemy hands," he explained.

Another body of Royal Marines which performed good work in Norway was the Fortress Unit, whose normal functions were the landing, transport and installation of anti-aircraft and naval coastal defence guns, and the construction of hutment camps, water storage tanks, pipe-lines and anti-tank obstructions, and the unloading of ships. During the winter of 1939 the Fortress Unit had done invaluable work in the United Kingdom, and on 11th May, commanded by Lieutenant-Colonel (later Major General) H. R. Lambert, D.S.C., it arrived in the neighbourhood of Narvik for the purpose of unloading coastal guns ammunition, oil and stores in preparation for the capture of the town. This was done, usually under heavy air bombardment, working in shifts all day and all night. On 23rd May a party of two officers and four Marines, under Captain H. G. Hasler, was detached with two motor landing craft to transport French tanks for the final assault on Narvik; it was necessary to take them by sea because none of the local bridges or car-ferries would bear their weight.

"Rearguard at Narvik"

The Narvik Peninsula was then entirely in the hands of the Germans, who were supplied mainly by seaplane and parachute. Against them were the French Foreign Legion, a battalion of Norwegian Infantry, and a force of Poles. All the British troops, with the exception of some artillery, had been withdrawn from the combat zone. Intercommunication was chiefly by Norwegian "puffers", heavily-built fishing boats, with a single-cylinder oil engine, which, as an officer remarked, is "simple and reliable but a most unpleasant shipmate in every way".

The Marines in the motor landing craft used the puffers as depot ships, eating and sleeping entirely on deck. On one occasion, an officer took an armed party ashore with the intention of shooting a sheep to supplement the monotonous rations, but returned empty-handed. The scheme had already been worked to death by the French Legion, and all he found was one sheep (evidently the last in the area) closely guarded by its owner, who also had a gun.

A squadron under the command of Admiral of the Fleet Lord Cork and Orrery opened the bombardment of Narvik on 27th May. The motor landing craft landed the tanks and the subsequent assault was successful, the enemy being forced out of the town into the mountains. By that time the remainder of the Royal Marine Fortress Unit had left for England, but Captain Hasler's party remained to assist in the evacuation of Narvik, which became necessary owing to the increasing bombing attacks on shipping, and helped to embark the rearguard, using a trawler, H.M.S. Man-o'-War, as a depot ship. She and two other trawlers were the last British ships to sail from Narvik, with the Marines on board, and thus, as the detachments of the Sheffield and Glasgow had been the first to land in Norway, Captain Hasler and his men were the last to leave.

CHAPTER 5

H.M.S. RESOLUTION'S MARINES

IN AE FIORD

At 3.30 a.m. on 9th May 1940, during the operations against Narvik, when H.M.S. Resolution (Captain, later Rear-Admiral O. Bevir) was lying off Tjelsund, the Northern Spray (Lieutenant-Commander D. J. B. Jewitt, R.N.), an anti-submarine trawler, came alongside and asked permission to discharge wounded. While on patrol she had sighted a German troop-carrying aircraft, which had been damaged by one of the Ark Royal's skuas, making a forced landing in Ae Fiord. A party of two officers and twelve ratings from the Northern Spray and her sister ship, the Northern Gem, had set out to investigate in a Norwegian puffer. They had seen a party encamped on the cliffs of the fiord, closed the shore and fired on them from the sea. The Germans returned the fire, with the result that two officers and five men from the trawlers were missing, and six wounded. Some of the missing men were believed to have been taken prisoner. The survivors had returned to the Northern Spray.

At 4 a.m. Captain Bevir called Major G. V. Walton, the Senior Royal Marine Officer, to his cabin for a conference with Lieutenant-Commander Jewitt. It was decided to land the whole of the Resolution's Royal Marine detachment, with a working party of four ratings, and two naval signalmen, to round up the German troops and to rescue the British prisoners of war.

The landing party (three officers and about 100 men, with a machine-gun section, commanded by Major Walton), embarked in the Northern Spray at 7.30 a.m. They carried 24 hours' emergency rations in canvas bags, besides tea, sugar and milk in bulk. A Lieutenant of the Royal Norwegian Navy, accompanied the party as interpreter.

At Ladogen, on the western entrance to Tjelsund Fiord, the party picked up the Northern Gem and a Norwegian pilot. The German aircraft had come down on the southern side of Tepkilnisset promontory in Ae Fiord, and Major Walton decided to land on the northern side of the hill, which was about 1,000 feet high, with the object of taking the enemy with an assault from the rear of their positions.

Since it was doubtful whether the trawlers could go up the fiord, two puffers were procured, and the landing party embarked in them, towing the trawlers' four lifeboats.

The Force reached the northern side of the promontory at 2.30 p.m. The water was too shallow for the puffers to go right in, so the trawlers' boats ran a shuttle service to the shore and landed the party without opposition. Major Walton then sent one platoon to advance along either bank of a stream which ran down the hill, and another to work its way along the coast to Tepkilnet village. He himself went ahead with two scouts. There was heavy snow on the ground and the rate of advance was under one mile an hour.

On the crest of the hill immediately above the village, three open camps were found. The first contained British equipment, the other two German. They had evidently been abandoned in haste, for personal kit, four Tommy guns, ammunition and food were lying scattered in all directions. A British steel helmet had "S.O.S." scrawled upon it in chalk.

The Marines continued their advance and swept down the steep hillside towards the aircraft. It was lying on the rocky foreshore, with one wing submerged in shallow water, the other buckled against the cliff. It was a Dorner 26, a type of civil flying-boat used by the Lufthansa in peace time on the South Atlantic service, and converted for troop-carrying. The position of the bullet holes showed that it had been shot down from above. After the guns, ammunition and documents had been taken from the aircraft, the Force re-embarked in the puffers, which had rounded the promontory with the boats and were lying in the little bay below the hill.

"Skirmish at Undereulet"

All this time there had been no sign of the enemy, but the Norwegian officer suggested that the prisoners might have been taken to Soetran, a hamlet of half a dozen houses about six miles farther up Ae Fiord. Escorted by the Northern Spray, the puffers accordingly proceeded to Soetran, where Lieutenant A. D. Comyn landed with his platoon and searched the village. It was empty, but showed signs of hasty evacuation. The platoon re-embarked and the Force returned to the head of the fiord.

Major Walton then received information from Resolution that the Germans were at Skjellesvik, a village lying to the south of Soetran and connected with it by a half-made road. The Norwegian Lieutenant and the pilot landed to confirm the information by telephoning to Skjellesvik and the neighbouring villages, but reported that they could obtain no news of the enemy. Nevertheless, Major Walton decided to investigate. The puffers were sent back and the Force proceeded in the Northern Spray.

It was a devious course by sea and it was midnight by the time they reached Skjellesvik. The inhabitants came out to meet the trawler in their boats (in those high latitudes there was no darkness at that time of year) and reported that the Germans and their prisoners were at Undereulet, a small village in a bay two miles to the north. There was no road between the two villages, and the bay was enclosed by steep cliffs. Rather than make a frontal landing by boat, however, Major Walton decided to send Lieutenant J. L. Carter with his platoon inland along the road to Soetran, with orders to swing to the left behind a saddleback hill and take the village from the rear. He himself made his way along the beach with the remainder of the detachment. The Northern Spray was to close the beach and give covering fire.

Some of the villagers volunteered to act as guides, but it was a hazardous and difficult march. The frontage was no more than ten yards, with sheer cliffs on the one side and deep water on the other. The Marines had to advance in single file, with no room to manoeuvre, and the rocky shore made the going slow. It took over two and a half hours to cover the two miles.

As the Marines approached, the Germans ran out of the house in which they were living and took up a defensive position at the base of the hill behind the village. They were driven out by the Northern Spray's 4-inch gun and the detachments' machine-guns, manned by the ratings and some of the Marines under Sergeant Coulson. This undoubtedly saved Major Walton's party from attack as they were forming up. The Marines then charged through the village with bayonets fixed. The Germans did not await their assault, but broke and retreated up the hill, leaving behind Tommy guns, rifles, bombs and ammunition.

The Marines surrounded another house, which appeared to be occupied by the enemy. The doors were locked, so they began to assault through the windows. They were about to get to work with their bayonets when they were hailed by a Cockney voice and discovered that the occupants were the prisoners from the Northern Spray dressed in German uniforms. The wounded pilot of the aircraft was also in the house, lying on a bed. He and the rescued men were sent back to the Northern Spray, while the Marines pursued the retreating Germans, supported by gunfire from the trawler.

"Round-up"

Since Lieutenant Carter's platoon might appear on the crest of the hill at any moment, Major Walton stopped the pursuit and ordered "Cease fire," content to leave the enemy to his detached party.

Meanwhile, Lieutenant Carter's platoon, accompanied by the Norwegian Lieutenant, marching towards Soetran, had reached an empty wooden building which had been used to house the road labourers. It contained bunks, a kitchen and a telephone. It was then 3.30 a.m. – broad daylight. The platoon had been marching through heavy snow – it had covered only five miles in three hours – and Lieutenant Carter could see that without skis it would be impossible for his men to cross the deep virgin snow to the hill over which the enemy was expected. He realised that if the Germans did appear they would make for the road. Therefore, having posted sentries, he occupied the house and let his men have a meal. An hour later the sentries reported the enemy crossing the saddle-back. They were making for the house, doubtless attracted by the smoke.

The Marines advanced upon them through the fir trees, firing as they went. In a few minutes ten of the Germans flung down their rifles and held up their hands. Two lay wounded in the snow. The prisoners were searched, relieved of their stick grenades and pistols, and taken back to the house, the wounded being brought in on bunks from the workmen's quarters. They proved to be troops from the Jaeger Battalion and had been on their way to Narvik when the Dornier was shot down.

Lieutenant Carter took the party to Soetran, the prisoners carrying their wounded. At 10 a.m. they reached the village, where they were picked up by a puffer and joined the remainder of the party on board the Resolution at 5 p.m., after an expedition which had lasted 34 hours. There had been no time for sleep and the Marines returned tired out but elated, with the satisfaction of having accomplished their task with complete success.

CHAPTER 6

THE LANDINGS IN HOLLAND AND FRANCE

"INSTANT DECISIONS: FOR FEAR FEEDS ON DELAY"

The Royal Marines' function of providing an emergency landing party from a warship for a special purpose on shore may also be performed by any of the three divisions, which hold themselves in readiness for sending a self-contained fighting unit overseas at an hour's notice, to prepare and hold a bridgehead, cover a demolition party, and, if need be, to act as a rearguard.

Such a call was made upon the Chatham Division at 7 p.m. on 11th May 1940, the day after the German invasion of Holland had begun. When the "General Assembly" was sounded, every man in barracks promptly fell in on the parade ground. There were but about 300 of them, for during that fateful spring the calls upon the Corps had been very great and besides fulfilling the demands of the fleet, parties had been despatched to guard neighbouring airfields, to provide a Cabinet Guard in London, and to defend Admiralty buildings.

The first 200 men on parade that evening were told that they were required for "a defensive role in an unknown place". Among them were instructors, cooks, grooms, boats' screws, storemen, a dining-hall attendant, and a barber. Many of these men were pensioners who had returned to the Corps on the outbreak of war. The oldest of these was 54. Some were in khaki, some were in blue. There was also the King's Squad, which had just finished their full training and had passed for duty that morning. They were ordered to return to their quarters, pack enough gear in their kit bags for three days, and to parade 50 minutes later in fighting order.

The Force fell in again at 8 p.m. and was organised into three platoons. Major B. G. B. Mitchell was in command, with three officers. They embarked in a convoy of buses with three days' rations, four machine-guns and two Lewis guns, and 30,000 rounds of ammunition. The police had cleared the roads and after what must have been the fastest bus trip in history (the convoy charged through Canterbury at 35 miles an hour), they reached Dover at 10 p.m. There they embarked in two waiting destroyers, the Verity and the Venomous. Soon after midnight they were steaming out of Dover Harbour.

Even Major Mitchell had been unaware of the Force's destination until he reached Dover, where he received his orders from the Vice-Admiral. Once at sea, he allowed it to be known that they were bound for the Hook of Holland, where they were to hold the port so that further landings could take place unopposed.

"Whitsuntide at the Hook"

The destroyers reached the Hook at 5.15 a.m. on Whit Sunday, 12th May, escorted by three Blenheims. Machine-guns were immediately mounted for anti-aircraft protection, and one platoon was put out as a screen. The rest of the Force was employed in disembarking the stores. Ten minutes after the landing, a flight of ME 110's attacked the Blenheims. In the ensuing battle, one Messerschmitt and one Blenheim were shot down.

Major Mitchell then held a consultation with the Commandant of the Dutch Naval Force, and learned that 200 German parachute troops were occupying a wood less than three miles from the port. He disposed his men on a perimeter a mile and a half round the jetty, sharing the positions with Dutch troops, some of whom were Marines. He made his headquarters the stationmaster's office on the jetty.

By 8 o'clock, twelve hours after leaving Chatham, the Marines were settling down in a front line alongside allies whose language they could not understand. But Marines are good mixers and it was not long before all were on friendly terms. When Major Mitchell inspected his positions, he saw the unusual sight of elderly pensioners drinking milk and wearing Dutch soldiers' overcoats.

"In order to make this milk more interesting," he recorded, "A rum ration was issued later in the day."

Meanwhile, one of the platoons had been suffering from German snipers in a row of houses to their front. A party of eight Marines and eight Dutch soldiers rushed the houses and burned them down. By this time, the British ships and aircraft had withdrawn. "I felt rather like a mouse in a closed room with a large and hungry tom-cat," Major Mitchell observed.

The rest of the day was quiet, however, until 7 p.m., when there was an air raid along the whole length of the waterfront. Then three German troop-carriers were seen flying low and slow over the wood and dropping more parachutists. As the parachutists emerged, they looked like small white butterflies. Then there were 30 white puffs, like bursts of shrapnel, as the parachutes opened. One of the Royal Marine machine-gun sections engaged them.

Early next morning, a composite battalion of Irish and Welsh Guards arrived from England and moved into the town. Within ten minutes of their landing, German aircraft were reconnoitring overhead and later a mechanised column was reported to be advancing on the town from the direction of Rotterdam. The Force stood-to, but the Germans halted a mile and a half away and dug themselves in, assisted by Dutch fifth columnists.

At noon the Force saw the Queen of the Netherlands embark in the destroyer Hereward, which sailed for England, and that evening the Dutch Government and a large number of Dutch refugees followed her. At 6.30 p.m. the Germans made a heavy bombing raid on the jetty. Major Mitchell's headquarters were bombed and machine-gunned. The potatoes which the cook was preparing for supper went sky-high and the tins on the shelves were riddled with bullets. An ammunition dump alongside was set alight, but the fire was gallantly extinguished by Marine S. Glenn. "It was a miracle that the Marines suffered no casualties," wrote Major Mitchell, "Even though they had made the best of every available cover. The steadiness of all concerned, in a most terrifying scene of leaping flames, exploding ammunition, and a destroyer badly hit and blowing off steam, made one proud of them."

During that night it became evident that it would be impossible for the Guards to advance to The Hague, as first intended. But the Queen and the Government had sailed in safety and it was decided that the Hook Force should be withdrawn. The Marines were to act as a rearguard, with a small party of Guardsmen to strengthen the left flank. In the morning the Germans systematically raided the town, bombing and machine-gunning the fleeing civilians. The Guards embarked in two destroyers, and at 1 p.m., when they were clear of the harbour, Major Mitchell brought in the rearguard. The withdrawal was made difficult because the driver of one of the commandeered buses was found to be a pro-German. He had to be persuaded to carry out his orders with a revolver at his back. A Quartermaster-Sergeant of the Irish Guards is said to have performed this duty most efficiently. During the Guards' withdrawal, Major Mitchell had re-embarked his stores in H.M.S. Malcolm, so that there should be no delay when the outposts came in. The arrival of the last party coincided with the appearance of 13 German bombers. The Marines scrambled aboard the Malcolm with all speed. Then one of the officers found that he had left the bag containing his washing gear in the bus and asked permission to return for it. This was granted. The Marines manned the whaler falls and hoisted the whaler on the port side. The naval demolition party, cool and unhurried, made their final preparations and embarked. The Malcolm sailed and as she left the harbour those on board had the cheering sight of ten destroyers coming in at full speed to screen her.

The Force reached Dover at midnight and arrived back in Chatham Barracks at 4 a.m. on 15th May, "Very tired, but very happy and thankful to be alive". There had been no casualties and so far from having lost any equipment, they had returned with "one anti-tank rifle in excess".

Major Mitchell was awarded the D.S.C. and two of his men the D.S.M. He was also decorated by the Queen of the Netherlands with the Order of Orange Nassau. Colour-Sergeant E. W. Parker was given the silver medal of the same Order.

"In Defence of Calais"

Similar "scratch companies" from Chatham were sent to Boulogne and Calais during those anxious days, to give cover to the naval demolition parties. The company sent to Boulogne, under

Serving in the Fleet. While the Royal Marine detachments in battleships, aircraft-carriers and cruisers share the many duties with the seamen, they are always ready to fulfil their special function of landing as an infantry force. Here they are seen on their own messdeck, which they call "The Barracks."

Anti-Aircraft Fort. In the Thames Estuary and along the East Coast, forts such as this defend the convoys against minelaying aircraft. Marines man the armament, mounted on the steel superstructure connecting the concrete towers, in which are the messdecks and magazines. Below, marines from the aircraft-carrier Victorious in Iceland.

Major C. F. L. Holford, on 23rd May, returned without casualties and was among the last to leave, having acted as a rearguard to the Guards, but the Calais party was less fortunate. Of the original strength of four officers and 81 other ranks, commanded by Captain G. W. A. Courtice, only 21 N.C.O.'s and Marines returned, and all the officers were lost.

They arrived at Calais in H.M.S. Verity soon after midnight on 25th May and first helped to defend the citadel, fighting alongside the French Marines. In the afternoon, one platoon and the machine-gun section were ordered to reinforce the Army in the front line. The machine-gun section was not seen again.

An hour later the citadel was bombed and shelled so heavily that it became untenable, and Captain Courtice ordered his men to withdraw to the jetty. They had to run across the blazing drawbridge and make their way through the town under a hail of bombs, artillery fire and bullets from fifth-columnist snipers hidden in the houses. Their numbers sadly thinned, they eventually took up a position on a ridge of sand dunes outside the town with the men of the Rifle Brigade, whose withdrawal they were to cover that night. The ridge was heavily bombed by German aircraft. Later Captain Courtice re-organised this position into double sentry posts with a frontage of 200 yards. The Marines held these posts that night and the following day. Sergeants Mitchell and East were despatched to collect any Marines who had become separated from the main party during the withdrawal from the citadel. When they returned on the morning of the 27th, the ridge had been vacated and there was no sign of Captain Courtice and his men. There is little doubt that they had become prisoners of war.

By 10 a.m. the position had become desperate. The German gunners could see every movement of the troops in the town. The German bombers renewed their attacks, unopposed. The two Sergeants then helped some of the wounded to embark in a small motorboat, the Condor, which had come in, but she ran aground during an air raid and had to be temporarily abandoned. Sergeant Mitchell brought in a badly wounded officer of the Rifle Brigade and decided to get hold of some rafts, carley floats and small boats that were lying outside in the harbour. Corporal Sowden and Marines Goodall and Beckham volunteered to swim out and bring them in. They were in the water for over four hours, coming and going. An M.T.B. picked them up, exhausted, as they were reaching the last boats. There is a story that as the M.T.B. approached them, Corporal Sowden looked up from the water and said, "Got any room? If not, I'll carry on."

The M.T.B. reached the inner harbour and took off all the Marines and soldiers present, except Sergeant Mitchell, who remained behind with the Captain and crew of the Condor in the hope of collecting more wounded men. At 4 p.m. two large bombs fell near the Condor. The waves caused by the explosions lifted her off the mud and refloated her. She was taken alongside at once, embarked a few more men, including Sergeant Mitchell, and surviving the attacks of the German bombers and a torpedo from a U-boat, reached Dover at 5.30 p.m. For their devotion in saving the wounded, Sergean Mitchell was awarded the C.G.M., Sergeant East the D.S.M.

Among the Marines who became prisoners of war was Marine S. F. Smith. While he was in the hands of the Germans he saw another prisoner shot against a church wall for attempting to escape. He saw other prisoners whipped because they could no longer march. He saw men bayoneted for accepting food from the French villagers. One night in Belgium he succeeded in escaping from his guards. Marching with feet bare and bleeding, unshaven, his ragged uniform covered with mud, and living on such food as he could pick up in the fields, he reached Brussels, hoping to get to Ostend and steal a boat, but only finding it impossible to make the coast. Befriended by a Belgian family, he then set off for France and eventually, after extraordinary privations and adventures, he reaches Marseilles, where he was imprisoned by the French, escaped but was recaptured within 500 yards of the Spanish frontier and was subsequently repatriated as unfit for active service. The record of his experiences was a shining example of that self-reliance in overcoming even the most appalling difficulties which seems the birthright of the Royal Marine.

LANDING PARTY

CHAPTER 7

REARGUARD IN CRETE

"SEEING CLEARLY WHAT HAS TO BE DONE" – THEN DOING IT

There is no better example of the versatility of the Royal Marines than what was the Mobile Naval Base Defence Organisation, which was more conveniently known as M.N.B.D.O.

This embodied what was a comparatively modern conception, dating back to 1923. The function of the M.N.B.D.O. was to provide the fleet with a base in any part of the world, whether on the coast of a mainland or an island, within a week, and then to defend it when prepared. The whole unit had at one point a combined strength of about 8,000, with a Major-General of Royal Marines in command. It was a body of specialists, including engineers and mechanics, transport and crane drivers; armourers and gunners; surveyors and draughtsmen, bricklayers, masons, carpenters, plumbers, painters, decorators and camouflage modellers; miners, blacksmiths, tinsmiths; even divers.

The M.N.B.D.O. was carried in specially equipped merchant vessels. The Landing and Maintenance Group was responsible for the collection of the required material, for putting it ashore in landing craft, and for transporting it once it was landed. The Group then completed its function by building wharves or converting existing jetties, making roadways from the beach, and erecting such buildings as was necessary.

The defence side of the organisation was divided into artillery groups with naval coastal guns, anti-aircraft and anti-tank guns, and searchlights to co-operate with all three. Since the organisation was also responsible for the general security of the base, it had a Land Defence Force consisting of rifle companies, machine-gun sections and light artillery batteries. The Navy provided the boom defences for the base.

Group Headquarters controlled the coastal artillery, the light and heavy anti-aircraft guns, the searchlights, the ordnance and workshop units and the sections responsible for the signals, postal meteorological, camouflage, provost, decontamination and medical services.

The whole organisation was designed in peace time with a nucleus of continuous service Marines or full time regulars you could say, to be expanded on the outbreak of war by the enlistment of special tradesmen and ordinary recruits, who after basic training were then moved to continuation training at the Technical Training Depot at Fort Cumberland.

If ever there was a place to induce those of a nervous disposition into going A.W.O.L. it could only be the 'Fort'. The massive walls, the portals, the earthy mounds and subterranean cell and prison dungeons, the endless stone and steel corridors, not to mention the reported sightings of shadowy figures in French Naval ragged uniforms (prisoners of yesteryear!), roaming the cellars after sunset when the flag was down and the silver bugle locked away for the night, was enough to push most people into learning their trade, passing their tests and getting as far away as they could from the legendary but black cold and ancien régime of Fort Cumberland.

The Corps expected a regular Marine artilleryman to understand any gun within a week. The Hostilities Only gunners were trained in the various divisions and also at Army centres. The anti-aircraft and searchlight units were employed in the Battle of Britain. One of the Royal Marine batteries, commanded by Major C. M. Sergeant, destroyed 44 enemy aircraft in 41 days and had the record bag for any battery in England.

"The Marines at Canea"

There were later two of these extremely complex organisations, with almost every man a highly-trained and combatant specialist. Early in April 1941, the first of them M.N.B.D.O. (1), under Major-General E. C. Weston, was sent to Alexandria expecting to be employed with the Mediterranean fleet, but after the evacuation of Greece had become necessary, it was decided to use it for providing a Naval base in Crete. Major-General Weston took command of the Suda area when Major-General (later Sir) B. C. Freyberg, V.C., was appointed Commander-in-Chief of the forces in the island.

Unfortunately, the decision to send M.N.B.D.O. (1) to Crete was not made in time for it to fulfil its complete function and only the advanced groups arrived shortly before the invasion began: the Landing and Maintenance Group, the Transport Company, the Signals Company, the Searchlight Regiment and one of the anti-aircraft regiments, some 2,200 men in all.

There were then about 28,000 troops in Crete, most of whom had been brought back from Greece, with a few Naval aircraft and R.A.F. Hurricanes. The area for accommodating these troops was restricted; there was no camp equipment; water points were few. Transport was almost unobtainable and the distribution of stores extremely difficult. All arrangements had to be improvised in the face of the imminent invasion of the island.

Australian and New Zealand troops formed the main body in the Suda area, which included the important Maleme Airfield, Canea, a town of some 1,600 inhabitants, with a good harbour and landing stage, and Suda Bay, the Naval base.

Shortly before the invasion began, Major-General Freyberg moved his headquarters from Canea, and an officer of the Royal Marines, who was the Provost Marshall on the staff of M.N.B.D.O. (1), was given command of the garrison. On 20th May, when the airborne attack began, the garrison consisted of a composite force of four officers and 300 men, including Royal Marines, Australians, and New Zealanders, some men of the Staffordshire Regiment, the Royal Army Ordnance Corps, the Corps of Military Police, survivors from the gun crews of merchant ships, one officer and three N.C.O.'s of N.A.A.F.I. and about twenty men serving "detention".

These troops were available for the southern defence of the town. The enemy was expected to attack with airborne troops from the south, then to land seaborne troops in the harbour. This appreciation of the situation proved correct. The Mediterranean fleet defeated the seaborne attack and, so far as is known, the only survivors of the invasion force were 14 soldiers and one medical officer who reached the shore on a raft. The first contact the Canea garrison had with the enemy was at 8.30 a.m. on 20th May, when a troop-carrying glider crash-landed on some rough ground outside the town. The garrison troops held their fire until the Germans had climbed out of the glider, then opened at 70 yards and killed the entire party of nine. Three more who had been killed on impact were found inside the aircraft. The garrison was able to salvage 11 Tommy guns and 1,000 rounds of ammunition, 60 hand grenades, two wireless sets, and some cameras.

The Germans continued to attack persistently with airborne troops, their method of landing gliders peculiarly reckless. They appeared to disregard the suitability of the ground and to take no thought of getting the carriers away so long as they could land the troops without losing too many in the resulting crash. The garrison dealt with the survivors so successfully, however, that only on three occasions did any penetrate into the town. The largest of these parties was three strong. The official report mentions that "they succeeded in doing a little sniping before being dealt with by the troops".

Many brave deeds were performed by the Royal Marines in the defence of Canea. Two may be quoted here. When a party of parachutists had secured a footing in a gun position, Captain A. L. Laxton made a determined attempt to get within grenade range, but was wounded in the head and back and in both legs. Nevertheless, he remained in observation and signalled back the numbers and position of the enemy, thereby enabling the gun position to be completely cleared.

On the second day of the invasion, Marine B. V. Jones, who belonged to one of the M.N.B.D.O.'s anti-aircraft batteries, shot down a German aircraft by firing a Bren gun from the shoulder. Later in the day he made a lone bayonet charge against seven Germans armed with sub-machine guns, who had landed in a glider close to his gun position. He was subsequently awarded the D.C.M. Captain Laxton received the M.C. The Royal Marine Signals Company, which was bivouacking in an orange grove outside Canea, also rendered valuable service by wiring for the anti-aircraft batteries, working by night.

"Defending the Suda-Maleme Road"

One of the Canea garrison's duties was to keep open the road between Suda and Maleme, the only supply line of the troops in the western sector of the island. By day, this road was subjected to intense air attack and had to be constantly cleared of debris and unexploded bombs. At dusk the garrison reported on the condition of the road, met the supply convoy from Suda, and escorted it to its destination.

"This operation was always an interesting one," declared the Royal Marine Officer in Command, "as the enemy was very clever at infiltration and there was usually a machine-gun post with fixed lines of fire along the road. A section of our troops went forward in two trucks to discover the enemy's position by drawing his fire; we then got round and attacked from the flank. These mélées invariably finished up with our troops using their rifle-butts, and I think they enjoyed these shows more than anything else. Having dealt with the position in this manner, the convoy was taken on until halted by another enemy post. The same procedure followed, until we delivered the goods."

On one occasion a convoy escort found a couple of Marines who had been manning a Bofors battery. The rest of the gun-crews had been killed and the ammunition was exhausted, but there had been no-one to give the survivors orders to retire. So they had remained in the slit trenches they had dug, defending their guns with their rifles. The Germans continued to drop more and more parachutists, and on the 27th they bombarded Canea from the air and demolished it. The garrison had sent away the civilian population on the previous night. The town was reduced to a heap of shattered smouldering ruins and since it was impossible to find food or clean water for the troops, they were ordered back to the Suda area. The Commander of the garrison, who had been wounded by a bomb splinter on the first day, now became a stretcher case after a hand-to-hand encounter with the enemy and was compelled to hand over his command to an Army officer.

"Covering the Withdrawal"

When the order came for withdrawal to the south of the island, Major R. Garrett, R.M., formed a number of anti-aircraft and searchlight units of the M.N.B.D.O. into a rifle battalion, which acted as part of the rearguard during the later stages of the evacuation. With some 4-inch and 6-inch guns and light anti-aircraft batteries, these 700 Marines kept the Germans at bay while the main body withdrew from the Suda Bay area to Sphakia in the south. They were told to fight until dawn on 1st June. This they did, with such effect that they made a further two days' evacuation possible, and so helped to save 17,000 troops to fight again. The evacuation was carried out by Major-General Weston, who assumed command of the troops in Crete on 30th May, when Major-General Freyberg was ordered to return to Egypt.

The losses of the rearguard were severe, and it was not possible to take off all the survivors. Once again, however, the Royal Marines' initiative and powers of improvisation rose to an emergency. One officer discovered a boat, and taking a mixed party of 60 survivors with him set out for the North African coast. Food ran out on the sixth day, the last rations issued being a lump of margarine dipped in cocoa. Now then, margarine dipped in cocoa.

In 1974, deep in a forest on the Irish border, I caught, skinned and cooked (burned), a rabbit basted with Army ration pack margarine in a cooking hole dug into the side of a small slope. What a total utter and complete disaster. Masters of the military theatre we may be, but for me, lumps of margarine dipped in cocoa would be my best bet and I must remember that in case they call me up again when I'm 72!

Now I diversify again, so to get back to North Africa, on the eighth day, during Divine Service, the party made a landfall and finally got ashore in the Sidi Barrani area.

Another party, consisting of five officers and 134 other ranks, including Naval ratings, Australians, New Zealand and Special Service troops, and 56 Marines, led by Major Garrett, put to sea in an abandoned motor landing craft. She was not designed to make anything but short trips at slow speed and her normal accommodation was 100 men. One of her propellers had been fouled by wire. The party put up at Gavdopula, a small island 20 miles south of Crete, where they cleared the wire from the propeller and filled up every receptacle they had that would hold water. Then they set off again to the south.

After the vessel had covered a further 80 miles the petrol gave out. An unsuccessful attempt was made to run the engines on fuel supplied for cooking. They then rigged a jury-mast and made sail with blankets, which one of the Marines secured with spunyarn – a mainsail of six and a jib of four. Since the unwieldy craft had no keel, she proved so difficult to steer that relays of six men had to swim alongside and push her head in the required direction. Major Garrett cut down rations to one-third of a pint of water, half a biscuit, and a cube of bully beef a day. No heart disease in the Royal Marines, so long as you stay away from the margarine and cocoa starters.

The above restricted ration allowance was issued by Colour-Sergeant C. A. Dean, R.M., who combined the offices of Purser and Master-at-Arms. His fellow Colour-Sergeant, H. C. Colwill, organised the watches and led the swimming parties.

Two of the Marines improvised a distilling plant consisting of petrol tins connected by a rubber tube and, by using the fuel which had failed in the engines, produced 4. x gallons of drinkable water in two days. Although two of the company died on the eighth day, the ingenuity of these Marines probably saved the lives of their remaining comrades.

Finally, at 1.30 a.m. on 9th June, having covered 200 miles, they too landed near Sidi Barrani. Two of the Maoris on board found water close to the shore. Sergeant Bowden, R.M., after reconnoitring five miles across the desert in the darkness, made contact with an anti-aircraft regiment, and arranged for lorries to be sent next morning. Thus, thanks to the initiative of two Royal Marine officers, 200 troops and two valuable Naval craft were saved from falling into the enemy's hands.

Nevertheless, many of the Royal Marine rearguard had no opportunity to leave the island when their work was done. Of the 2,200 who had gone to Crete with the M.N.B.D.O., only 1,000 returned.

As with the four Marines of the Wager, whom R. L. Stevenson, in his essay "The English Admirals", enshrined in the traditions of the corps, "There was no room for these brave fellows in the boat," and if those left behind in Crete did not stand upon the beach at Sphakia, give three cheers and cry "God bless the King," like those others upon the beach of that lonely island in the South Pacific, there is little doubt that they, too, "made it a point of self-respect to give their lives handsomely." Some determined to give them so handsomely that when a Sunderland arrived to take off survivors, they declined to go. They made no fine speeches; it seems that they had no feelings for that piece of purple which, in the estimation of R. L. S., embellishes noble action. Gathering up the rations and the ammunition the Sunderland had brought, they turned their backs upon safety and security without more ado and retired into the hills to carry on the fight. There, it would appear, they fought on waiting until a later date to welcome their comrades from the sea. For if the Royal Marines were among the last to leave Crete, we may be very sure that they were among the first to return.

CHAPTER 8

SINGAPORE

"HERE FOUGHT THE PLYMOUTH ARGYLLS"

After the survivors from M.N.B.D.O.(1) had returned to Alexandria from Crete, the unit was re-formed. The Land Defence Force remained in Egypt, waiting for deployment; other units were sent to the Maldive Islands and Ceylon.

Meanwhile, the war further east was going badly. Marines fought gallantly in Hong Kong before the capitulation. I will later deal with my views on the catastrophic failure to inflict albeit guerrilla warfare on Japanese infantry not on the island of Hong Kong itself, but across the fragrant harbour into Kowloon and then on to the ideal theatre of the New Territories where a well-trained band of brothers with experienced Royal Marine officers should, and could, have caused frustrating mayhem to the invaders. Then, on 10th December 1941 came the sinking of the Prince of Wales and the Repulse. The survivors from the Royal Marine detachments, about 300, returned to Singapore. Some went to hospital, the remainder were re-kitted at the naval barracks and were organised as a battalion for duties ashore, under Captain R.G.S. Lane, who had commanded the Repulse detachment. Captain C.D.L. Aylwin, of the Prince of Wales, took one party to guard the wireless station at Kranji. Another was given a course of training in jungle warfare and went to Northern Malaya for the purpose of attacking the Japanese behind their lines, led by Lieutenant R.J.L. Davis, of the Repulse. A third was sent up-country to catch the fifth columnists who were known to be signalling to the enemy by night. At the same time, the Marines who had disembarked from the other ships, and had a knowledge of the inland waterways, joined the naval ratings in some small craft fitted with Lewis guns to patrol the Straits of Jahore.

Lieutenant Davis's party went by train to Kuala Lumpur, then to Port Swettenham, where they camped in the Chinese School. While waiting to begin the raids, they continued their jungle training. It must be remembered that these men had come straight from the battleships, and knew nothing of travelling in the jungle, let alone fighting in it.

I can assure you, from my own experience at jungle warfare school it is hot, exhaustive work cutting your way through virgin jungle with a good steel machete, so how these men did it with no more than bayonets can only be left to the imagination. The worse bugbear was the red ant – could he bite! The mangrove swamps were not too funny, either; if you had the misfortune to get into one, you promptly sank in up to your waist, surrounded by small red crabs, you take on a coating of foul-smelling slime, the aforesaid red ants - just like the Japanese at that point in the war, appeared to be able to walk on water - and just millions of mosquitoes.

Before the Marines could embark, Japanese aircraft machine-gunned and sank the motor launches in which they were to have made their raids. Having no water transport left, they were attached to army units, and after five days were forced to withdraw to Kuala Lumpur, where they eventually acted as a demolition party. "We were given a sledgehammer each," wrote one Marine, "And driven to five engineering sheds, with orders to destroy everything. We had a great afternoon. I don't think the Japanese got much value out of that machinery."

They also blew up ammunition dumps, buildings and ships in the harbour, and were the last troops to leave. All this time they had been living off the country – on bananas, pineapples, mangoes, coconuts, fowls and ducks. They were soon on the move again and came under constant air bombardment, which cost them their first casualties. They patrolled the river banks and held an important bridge to cover the withdrawal of two army battalions. Finally, they joined up with a convoy and returned to the naval base a few days before the Japanese reached Jahore.

The last British troops to cross the causeway which links Jahore with the island of Singapore were the 2nd Argyll and Southerland

Highlanders (Lieutenant-Colonel I.M. Stewart, D.S.O., O.B.E., M.C.). They came over with their pipes playing "Hieland Lassie" and blew up the causeway behind them. Their strength had been reduced to 250, and the Marines from the Prince of Wales and the Repulse were sent to join them at their hutted camp outside Singapore. The composite battalion thus formed was officially called the Marine Argyll Battalion, but it will forever go down in history as the Plymouth Argylls; the Prince of Wales and the Repulse had been "Plymouth ships". The Marines formed two companies, each with an armoured car and a Bren carrier, and the fellowship they created with the gallant 93rd will be an abiding memory in both Corps. "We never want better companions," said Colonel Stewart.

The Japanese attack on Singapore began in the very early hours of 9th February 1942. The Plymouth Argylls were despatched to Bukit Timah, four miles from the city. The Japanese were shelling the causeway, and the lorries were dive-bombed and machine-gunned continuously. It became necessary to abandon them when the Japanese artillery got the range of the road. The Royal Marines and the Highlanders marched through rubber plantations to their allotted positions, where they began to dig themselves in, suffering many casualties from the air raids. Blazing oil tanks lit up the sky and made it as bright as day. The tropical rain on that particular night poured down in sheets and filled the trenches, so that the troops spent half the night up to their waists in water, while covering the withdrawal of the Australians. One of the armoured cars manned by the Marines made periodical sorties to a point on the causeway where it could fire effectively on the Japanese working parties.

At 6 a.m. next morning, Marine R.W. Seddon was "Having a bit of a swill in a claypit," as he put it, when he heard the crack and thump of rifle fire. Captain Lang had been surrounded. Seddon thus described the incident: "Colonel Stewart cried, 'Come on, Marines,' and we charged forward with our Bren guns. The Japs wore all sorts of rigs. Some were in shorts, some with equipment, others without; some wore only sarongs. You couldn't tell whether they were Japs or Malays or Chinese. The undergrowth was very dense, and we had to open up at random. I was doing a bit of spraying with my Tommy gun and got some of them. Captain Lang had defended himself with his revolver and joined up with us. That night we had no sleep. The Japs must have been within 50 yards of us. We could hear them shouting in their own brand of English: 'Any British or Indian troops here! If so, come out, the war is over!' We went on a bit of a patrol next day and saw half a dozen strutting along. At the time we weren't sure what they were, so I challenged them. They answered, 'Punjabis.' After going a few paces further they suddenly opened on us, hitting an Argyll officer."

With Seddon was one of the boy-buglers from the Prince of Wales. He had proudly declared on numerous occasions he was "nearly fifteen". "On the night the Japs invaded the island everything went haywire," he said. "During the fighting a Jap sniper who had been hiding up a tree jumped down on me and wounded me in the wrist with his bayonet. I couldn't stick him myself, so I called the Sergeant, who finished him off."

"Last Days in Malaya"

The Plymouth Argylls had to operate in thick undergrowth, with no field of fire, their only protection in the air two army co-operation Lysanders. The Japanese bombers flew low and flung hand-grenades among the trees. But the Marines held up the advance as they withdrew and destroyed several tanks with their anti-tank guns, enabling the nursing sisters and patients to evacuate an army hospital, which the Japanese bombed and burnt out an hour later.

During the retirement the Marines became split up, and made their way back to the city in small groups, their rifles at the ready, for there were snipers all the way. They continued to lose many of their comrades from air attack. "There was no-one left to put up a show," said one of the survivors. "They were all squandered up."

On reaching their camp, they found that it had been turned into a hospital. Some made their way to Keppel Harbour, to be told that a general evacuation had been ordered. Some of them boarded ships and got away while Singapore was blazing. One party found a rowing boat and after pulling for four hours was picked up by the gunboat Grasshopper. She was dive-bombed and sunk off the Sumatran coast. The survivors swam ashore, found an abandoned motor bus, which one of the Marines managed to get going, then drove to the nearest town. There they went on board a destroyer and were taken to Batavia. The Dutch authorities mistook them for spies (for they were in rags by that time) and arrested them, but they were released later and eventually reached Colombo.

Yet another party left Singapore in a motor launch commanded by a Lieutenant, R.N.V.R. They reached the Dutch island of Singkep where they found 800 soldiers and seamen. Nearby was another island where there were a number of Australian nurses and white women refugees. The Japanese were systematically bombing it and sending boats to capture the women. Some of the Marines at Singkep joined a party of volunteers in a motor launch towing flat-bottomed boats, brought off the women who were still alive, and buried the dead. They finally left Singkep in a motor-boat with 300 on board, towing another, which sank. The survivors were taken aboard the first boat, and all reached Padang, in Sumatra, whence they, too, were sent to Colombo.

Such is the story of the Royal Marines in Malaya. It was a bitter period in a God-awful environment to be stuck with no promise of reasonable logistical back-up. I know this theatre well from a long period I served with 40 Commando and then 42 Commando in 1968/69, but do not put all the blame on the men on the spot, who at least did their very best in difficult circumstances which I firmly believe has never been properly researched or appreciated in this country. I went to enormous lengths with the M.O.D. and H.M.S.O. to find scraps of information which might penetrate the inner depths or even the peripheries on the thinking of the defence of Hong Kong and Singapore at senior command level, but in the end I could only envisage a number of Whitehall mandarins forever bleating to the Chief of the Imperial General Staff, "Don't argue, just do it."

The Royal Marines, and those with them, had fought, there is no doubt in my mind, a losing battle; a battle which had been lost before ever they had gone into action from their ships, unused to jungle warfare, utterly ill-equipped into this logistical green hell of a nightmare, but with their courage high.

Some day, it may be, a memorial will be raised to the Plymouth Argylls in Singapore Island. The inscription on it might well take the form of that which is placed above Nelson's quarters in Port Royal, Jamaica: "Here fought the Plymouth Argylls. Ye who tread their footsteps, remember their glory."

Sadly, even today there are historians and others who would scoff at their efforts in the face of monumentally superior forces which drives me to such anger as to quote, in situations like this, the words of a man who had tasted the true meaning of life on the edge of conflict: It is not the critic who counts nor the man who points out how the strong man stumbles or where the doer of deeds could have done better. The credit belongs to the man who is actually in the arena, whose face is marred by dust and sweat and blood, who knows great enthusiasm, great devotion, and the triumph of achievement, and who, at worst, if he fails at least fails while daring greatly, so that his place shall never be with those odd and timid souls who know neither victory nor defeat. You've never lived until you've almost died. For those who have had to fight for it, life has truly a flavour the protected shall never know.

Amen, to the members, one and all, of the Plymouth Argylls above all to those who will remain forever where they fell, for these men who tried to do something and failed are infinitely better than those timid souls indeed who tried to do nothing and succeeded. So, overboard with the critics say I, and for the brave few of the Plymouth Argylls let us remember:

"Cowards die many times before their death, the valiant never taste of death but once.

That's a valiant flea that dares to eat his breakfast on the lip of a lion." - William Shakespeare.

"Singapore: Simbang Revisited"

Before we "saddle up" and yomp away to the Irrawaddy and Burma, let me tell you about one of the finest land postings the Royal Marines had before the Union flag came down for the last time on the ashes of Empire. From 1959 to 1971 many Royal Marines were based in Singapore. The majority, 42 Commando and H.Q.3 Commando Brigade, in H.M.S. Simbang, which was in the Sembawang area of the north part of the island.

Dieppe Barracks was built alongside H.M.S. Simbang, on the other side of the airfield, in March 1966, to house 40 Commando, who moved from Burma Camp in Malaya. In a visit by Lt. Col. Andrew Noyes R.M. included a trip to the "old" H.M.S. Simbang, now the Singapore Air Force Sembawang Helicopter Base. In their small museum on the base the history of H.M.S. Simbang is displayed. The history of the base began in 1934/35 when the site, originally part of the Bukit Sembawang Rubber Estate, was purchased by the Air Ministry. Approval for the construction of a grass airfield was given in 1936, and the British Army started work the following year. Built for the R.A.F. and originally planned for two bomber squadrons, the airfield was transferred to Admiralty control in 1939.

In 1940 it was transferred to the R.A.A.F. and in August two Australian squadrons were established. The following year, two fighter squadrons were added to the strength. The war reached Singapore in earnest in December 1941 and the station suffered heavy raids and many casualties. By January 1942 the situation in Malaya had deteriorated to the extent that Sembawang

was evacuated. In mid-February it was occupied by the Japanese, but little is known about its use during this period.

After the Japanese surrender in September 1945, a naval advance party returned to take control. They found about 90 Zero fighters on the airfield and some 700 Japanese officers and men. The station was honeycombed with tunnels and foxholes and in a state of considerable disorder. Work on rehabilitation was started immediately and Japanese prisoners of war were employed filling in the foxholes and tunnels, and laying a 1,200 yard pierced steel planking runway. On 14th December 1945, the air station was commissioned as H.M.S. Simbang.

In January 1948 Simbang was handed over to the R.A.F., and in September 1949 was reduced to care and maintenance. It was recommissioned on 16th January 1950 when war again threatened in the Far East, and the workshops were built up to full aircraft repair yard standards with a throughput of about 20 aircraft per month.

The end of the Korean War overlapped the Malayan Emergency and in January 1953, Sembawang witnessed the operation of helicopters for the first time when 848 N.A.S. arrived to give transport support. In July 1953, the station task was reduced to supporting 848 Squadron aircraft from visiting carriers and a holding unit equipped with Sea Furies and Fireflies. Eventually the fixed wing aircraft returned to the U.K., and by September 1955, the reduction in complement had been completed. At the end of the Malayan Emergency in April 1957, H.M.S. Simbang was again reduced to care and maintenance.

Early in 1959, work began on extending the galleys and accommodation to transform Sembawang into a base for a Royal Marine Commando and a Naval Helicopter Squadron and the advance party of 42 Commando disembarked from H.M.S. Bulwark. During March 1961 H.Q.3 Commando Brigade arrived and at the end of July 1961, 848 Squadron and 42 Commando were disembarked onto Sembawang for three months.

H.M.S. Simbang commissioned for the third time on 4th September 1962 under the command of the Deputy Commander, 3 Commando Brigade. In late 1962 and early 1963, 814 and 815 N.A.S. spent periods disembarked there. The years 1963-66 saw H.M.S. Victorious Centaur, Eagle, Ark Royal and Hermes operating in the Far East, and in each case the attached anti-submarine squadrons spent time at Sembawang. Other activities at the air station included joint R.N., R.A.F., and army supply dropping practice, parachute dropping by students of the R.A.F., Far East parachute training school, and Commando exercises.

In 1966, a Fleet Amphibious Forces Base was established at H.M.S. Simbang, divided between Kangaw Barracks to the south and Dieppe Barracks to the north east, with the airfield in between. Dieppe Barracks was completed in March 1966 when 40 Commando moved in from Burma Camp and the First Commando Light Regiment Royal Artillery was established at Sembawang.

On 1st September 1971, administrative control of the Fleet Amphibious Forces Base (Far East) was handed over to the Anzuk Support Group H.Q. and H.M.S. Simbang paid off on 30th September.

HMS *Simbang* in the 1960s, showing the airfield and 42 Cdo barracks, with Dieppe Barracks to the left of the airfield in the middle distance.

(Right) HMS *Simbang* from the air in the 1960s

HQ 3 Commando Brigade and 42 Commando lines

"Hong Kong and Stonecutters Island"

I said I would come back and give you a brief indication of this island in the sun. When a colony comes under a particular flag, be it British French or whatever, the business interests of the country involved will always be a bugbear as far as the military are concerned, especially in the event of foreign attack. If you immediately surrender, the colony's business interests will possibly be preserved. On the other hand, if you plunge your military into a covert theatre of operation against the invaders to cause and maintain every possible form of aggravation, they will more often than not take retribution in many ways. For every invader you kill, perhaps ten of the civil population will be taken out, lined up and shot. This certainly happened with the Japanese in the Far East.

Wars are never glamorous, they are dirty grimy affairs and hugely dependent on control of the population. This in turn, causes friction between political decision-makers who are far too often driven by self-interest rather than the wisdom and experience of senior military commanders of vision and courage, of initiative and imagination, and above all a deep and intimate knowledge of the local terrain and its natives.

In an ideal world, before the Union flag was lowered in Hong Kong, the Japanese should have suffered much more, and for longer than they did. On the mainland opposite Hong Kong stands the city of Kowloon. Behind Kowloon is what was known as the New Territories, which was a piece of China on loan to the U.K., for arguably one hundred years. On this you can give or take ten years, as we were involved with the whole area since 1854.

Before involvement with the Royal Marines, I was stationed for a three-year period with the Royal Warwickshire Regiment. We were in an open hilltop camp close to the Chinese border by the village of Fan Ling. The only possessions the natives had or cared about were the paddy-fields, water-buffalo, chickens and ducks by the thousand, and dogs, mongrel dogs that roamed and barked incessantly day and night. The mosquitoes were so thick just before sundown every evening, they would blanket you if you did not use plenty of repellent - bug dope, as we called it.

Over those years we became very familiar with the whole area of the New Territories. At one point we had slit-trenches and hides on many of the hills and high peaks and pathways through the terrain that only we knew. The natives seldom travel to the hilltops except to certain areas to bury their dead in earthen urns, which they dig into the hillsides. It was very plain to the military way of thinking that with intimate local knowledge not only with the terrain but with the locals themselves, the New Territories from the back of Kowloon right up to the Chinese border was an ideal platform to harass the Japanese if only the will and initiative had been there. It makes me sad to think there are certainly officers and men around today who would have left the Japs in a whirlwind of frustration if only given the chance. Well, wrong politics, wrong time and place I guess! But there is no doubt in my mind with the right equipment, officers and men, as the flea drives the dog mad, we could, and should, have driven them to despair, for what did the Japs know of the Fan Ling Valley; from Cap Badge Hill to Birds Hill, through Birds Pass to Snake Pass, from there to the rocky shale of Lead-Mine Pass and up and away to the Jubilee Reservoir. Hellish terrain all the way in heat that could reach up to 95 degrees in mid-afternoon.

If only we knew then what we know now we might not look back in such anger, but before I cease this long and terrible digression and as our story involves the Royal Navy as well as the Marines, I think they deserve their Hong Kong involvement to be mentioned.

Snake farm, gunpowder magazine, prison, smallpox refuge, hospital, military base – Stonecutters Island has been many things to many people. No longer an island – the world's busiest harbour has "reached out" and engulfed one end of it – Stonecutters has provided a barracks for the final resident British battalions and a headquarters for the Hong Kong Military Service Corps.

Its buildings, a patchwork of ancient and modern, reflect the many purposes to which its now peaceful acres have been subjected even in the presence of an infantry battalion, the island remained a tranquil place. In its last days as a British territory, it appeared worlds away from the sky-scrapered city shimmering out of the harbour haze just a short ferry-ride away.

Named after the peasants who quarried its resources centuries ago, Stonecutters was claimed for the Crown in January 1861, when the Chinese Manchu Empire ceded it and Kowloon in perpetuity. As Hong Kong had been established in 1843, the island was immediately drawn into the colony. A gaol was built on Stonecutters in 1866 and a prison hulk anchored off its shores. Someone had a change of mind and the cells were never used for their intended purpose.

A powder magazine was started in 1870, a forerunner of what was to be the Royal Navy Armaments Depot. Subsequently, it became an Army depot, later known as the Ammunition Sub Depot R.A.O.C.

When smallpox threatened to decimate the colony in the 1870's, the Government attempted to isolate carriers of the dreaded disease by placing them in quarantine on Stonecutters. A legacy of the time is the grave on a grassy bank in the north west corner. It is the last resting place of William L. Anningson, an Inspector of Police who contracted smallpox and perished on the island.

While the bodies of Chinese patients were claimed by relatives, European victims were buried at sea in an attempt to contain the disease. Why Inspector Anningson was buried on the island is not known.

Before the turn of the century, a hospital for poor people was built, and gun batteries were constructed for the defence of the island. In 1980 it was designated a military reserve area and thereafter closed to the civilian population.

By the outbreak of the First World War, Stonecutters was predominantly Royal Navy, but with Royal Artillery personnel manning its gun batteries. Much of its use was recreational. A large Chinese population – mostly Amahs and boys working in the quarters, messes and barracks – moved in. Stallholders and small businesses opened up to do trade along Range Road, enabling residents to do most of their shopping on the island at less than half of Hong Kong prices.

Underground storage chambers for the Royal Naval Armaments Depot were completed in 1935, and the aptly named "Wuthering Heights" quarters, used for officers' accommodation, dates from that time. When the Japanese attacked in December 1941, troops stationed on Stonecutters withdrew by ferry after practically every building had been hit. The Japanese held it for the next 3 _ years, using it mostly for rest and recreation.

They also introduced a snake farm to produce and collect serum. Their medical teams needed this through many of their conquests around the Far East and apparently this programme was very successful.

There is an unconfirmed story that the Japanese in the immediate area of Hong Kong were among the best behaved in the Far East. It was said that their garrison commander when dealing with soldiers who were found guilty of misconduct, would award them with punishment of anything up to 28 days administration and maintenance of the snake serum farm which, after the first few days would reduce the average Jap Superman to a quivering wreck. Don't quote me on this as it's a well-known story I picked up in the bars of Wanchi, but it makes so much sense doesn't it! The Japanese social workers answer to the ne'er-do-wells.

Well, a gun mount on "Wuthering Heights" is Japanese, but today little other evidence of the occupation remains except, in comparison to the New Territories the wild and sun-drenched Stonecutters Island seems, they say, to have more than its fair share of shadowy atmospherics, weed-encircled rusting old helmets, but above all else, snakes.

CHAPTER 9

FORCE VIPER ON THE IRRAWADDY

Well, after that small diversion to the sun-drenched glories of Hong Kong, let us now get back to work on the closing stages of the campaign in Malaya, when the threat to Burma was becoming increasingly grave, the units of M.N.B.D.O.(1) in Colombo were asked to find volunteers for "special service of a hazardous nature". The response was immediate, and a party of four officers and 102 other ranks was selected, commanded by Major D. Johnston, with Captain H. Alexander as Second-in-Command and Surgeon-Lieutenant A. J. Innes, R.N.V.R., accompanying the Force as Medical Officer. It was known as Force Viper and left Colombo in H.M.S. Enterprise on 8th February 1942 for an unknown port.

Once at sea, the Force learned that its destination was Rangoon, where it disembarked on the 11th. The city had a derelict appearance. There were few people in the streets and piles of garbage everywhere. The British had lost Moulmein and were falling back towards the Sittang River. It was thus no longer possible for Force Viper to perform the role for which it had been formed – to patrol the east coast of the Gulf of Martaban in small craft in order to prevent the Japanese from getting behind the British troops – and Major Johnston was ordered to form a flotilla for other duties.

He acquired (that word appears frequently in the global initiatives of Her Majesty's Royals) a 35-foot diesel motor-boat, the Alguada, and began boat-training on the Rangoon River. Lieutenant W.G.S. Penman, of the Burma R.N.V.R., was attached to the Force. He was an experienced yachtsman and a qualified engineer, spoke Hindustani and Burmese, and his knowledge of local conditions was to stand the Force in good stead. On Major Johnston explaining the difficulty of obtaining boats for training purposes, he disappeared for a few days and returned with four Government touring launches – the Doris, Rita, Stella and Delta – and the diesel motor-boats Ngazin and Ngagyi, of the Irrawaddy Flotilla Company. Now, there was initiative on a scale fit for kings. Each launch was equipped with a Vickers machine-gun, an Aldis signalling lamp and a wireless telegraphy set. Bren guns and mortars were added later, as acquired. These launches formed the nucleus of the flotilla throughout the campaign.

Lieutenant Penman then took over the training of the boats' crews, when they could be spared from other duties, which include close defence of the oil refineries at Syriam, three miles from Rangoon; and, since all the police forces in the area were withdrawn on 22nd February, the Marines were given the task of maintaining order in the town. "In the racing tides of the Rangoon River, this boat-training was a tiring and at times, a hair-raising job of work," wrote Major Johnston in his official report, "But the men were first-class, well trained and very keen to learn all the tricks of staying alive, so that within a fortnight we were able to start river patrol work with reasonable efficiency and an unimpaired fleet."

By that time the Japanese had crossed the Sittang and fighting was in progress in the Pegu area, 50 miles north-east of Rangoon. In Rangoon itself the feeling of insecurity grew day by day. The question was no longer if, but when, the city would be evacuated. Arson and looting were on the increase and large fires were burning continuously both in Rangoon and Syriam. Major Johnston instituted car patrols and offenders were publicly flogged. Even so, the looting and the fires continued.

On 4th March, Major Johnston received orders to patrol the Rangoon and Pegu Rivers, the Twanti Canal and the Rangoon Docks. Looters were to be shot on sight. A party was to be left behind to assist in the demolition of the oil refineries. The Force was divided into three platoons, each taking one launch, with a motor-boat attached. The Delta was used as a store-ship and carried the demolition party. The motor-boats were manned entirely by Marines; the launches still had their Chittagonian crews, who remained loyal throughout the campaign. The Force had been given special permission to sail under the White Ensign.

"MANDALAY" - 1892 - RUDYARD KIPLING

By the old Moulmein Pagoda, lookin' lazy at the sea,
There's a Burma girl a-settin', and I know she thinks o' me;
For the wind is in the palm-trees, and the temple-bells they say:
'Come you back, you British soldier; come you back to Mandalay!'
Come you back to Mandalay,
Where the old Flotilla lay:
Can't you 'ear their paddles chunkin' from Rangoon to Mandalay?
On the road to Mandalay,
Where the flyin'-fishes play,
An' the dawn comes up like thunder outer China 'crost the Bay!

This poem sets the mood so well.

"The Clash at Henzada"

The military evacuation of Rangoon began on 6th March. The demolition of the refineries and the docks took place next day. Soon the sky was filled with immense pillars of black smoke. In one refinery alone, 20 million gallons of aviation spirit were destroyed. The flotilla covered the embarkation of the demolition parties, then made its way through the Twanti Canal and the China Bakir into the Irrawaddy. The Marines destroyed, or took in tow, all the power-boats they passed, and set on fire a large dredger.

"We steamed by day only," wrote Major Johnston. "Our routine was semi-naval. Reveillé at 0600. Scrub decks till 0645, by which time it was light enough to see the buoys, and we got under way. Then breakfast, and quarters to clean guns. Lookouts were on watch at all times. We went ashore at the towns to buy local produce and get what information we could. We had been lucky enough to procure 50 crates of beer before leaving Rangoon. Grog was issued after anchoring each night."

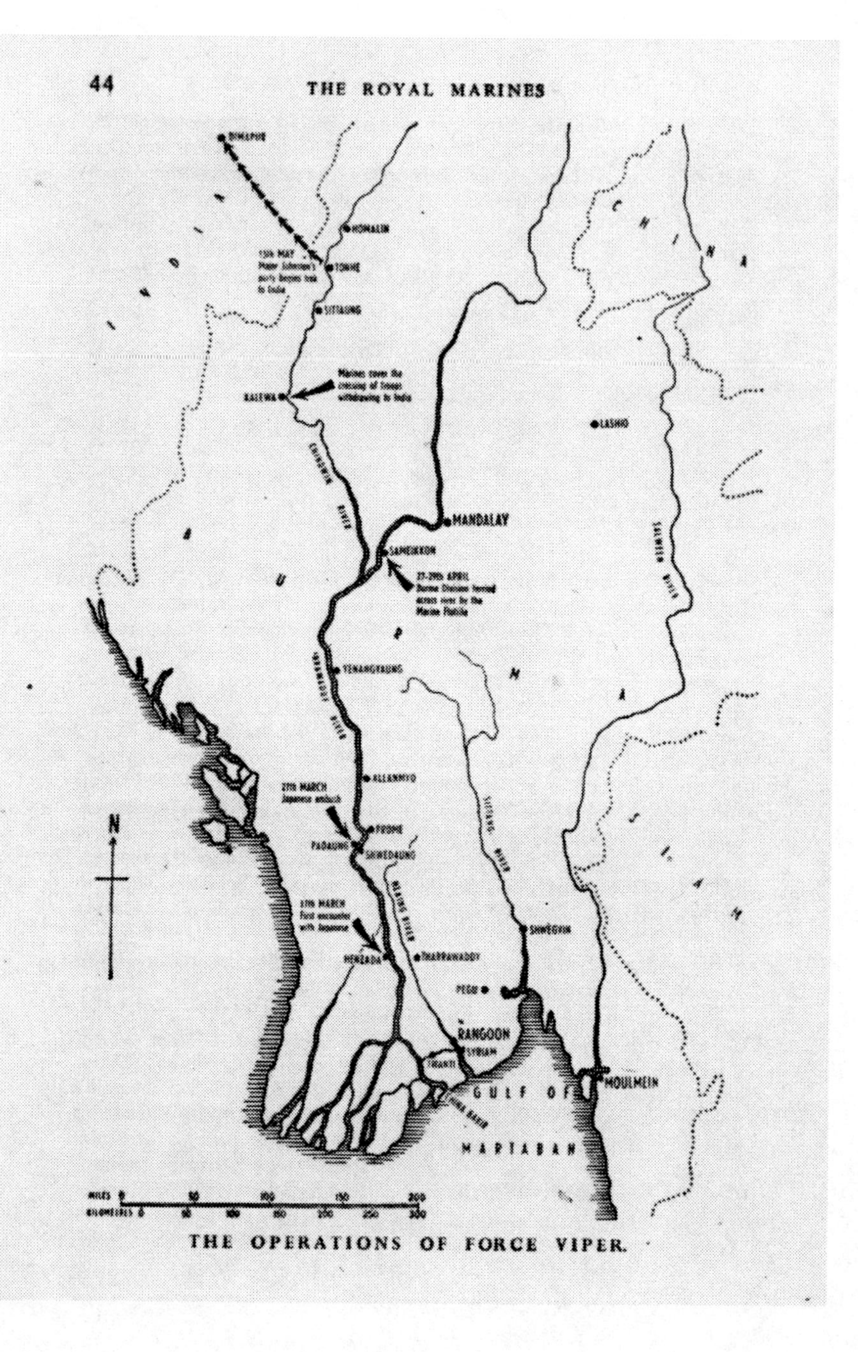

THE OPERATIONS OF FORCE VIPER.

The flotilla reached Prome on 13th March after a peaceful passage and was attached to the Seventeenth Division, which was in combat with the enemy in the area north of Tharrawaddy. Major Johnston's orders were to protect the Division's right flank by preventing the Japanese from coming up the river behind them and crossing from west to east. The store-ship was now replaced by the Cynthia, a 110-foot steam launch, the Ngazin and the Ngagyi by another diesel boat, the Snipe, and two armoured kerosene motorboats, the Xylia and the Delta Guard 9.

The first task of the flotilla was to assist Burma Commando II in carrying out demolition at Hensada, some miles below Prome. The Royal Marine demolition party was attached to the Commando, with the Rita as escort to the Hastings, an Irrawaddy flotilla double-decker. The two vessels reached Henzada on 17th March. The Hastings secured to the bank and the Rita lay off close to her. A small party from the Commando and Force Viper went ashore to reconnoitre. When they had advanced 200 yards towards the village, a Burmese civilian called upon them to surrender, saying that they were covered by machine-guns. "They gave an unprintable reply to this," observed Major Johnston. Japanese then appeared in large numbers and the party started to fight its way back to the river. Hearing the firing, the Rita pulled out into midstream and drew most of the Japanese fire, which she returned vigorously with her Vickers and five Brens, the guns "getting right in amongst the enemy". Mortar bombs began to fall around the ships and the Hastings pulled out from the bank, but went in twice to pick up the shore party. Mr. Rae, the Master of the Hastings, handled his ship from an exposed position for'ard with great coolness, and throughout the action the two Chittagonian Serangs remained at the wheel unperturbed.

Once the shore party was re-embarked, both vessels made off upstream without being hit by the mortars. The Rita was hit repeatedly by rifle-fire, however, but the Japanese bullets had poor penetrating power, being stopped by such unlikely articles as a rolled blanket or a tin of food. One Marine and two men from the Commando had been lost, and two men in the Rita were slightly wounded. Later information disclosed that the enemy lost over 100 killed.

This action showed Major Johnston that a force put ashore from the flotilla would always be at a disadvantage, and that his proper role was to destroy the enemy on the river. Even there, as he had come to realise, the flotilla had limitations. The launches were slow and restricted to the main channels. The Burmans were cutting away the buoys and removing the crossing marks on the banks. The flotilla could not move with any certainty at night to make a surprise attack and its operations must be mainly defensive. The patrols continued. The Marines were learning the river. At first the motor-boats had been like a pack of unruly hounds, dashing off to chase country boats up creeks and often ending by going aground, so that a party would have to go over the side to push off. Now great caution was needed, for the river was falling and the channels were becoming increasingly difficult.

"But that's all shove be'ind me – long ago an' fur away,
An' there ain't no 'buses runnin' from the Bank to Mandalay;
An' I'm learnin' 'ere in London what the ten-year soldier tells:
'If you've 'eard the East a-callin', you won't never 'eed naught else.'
No! you won't 'eed nothin' else
But them spicy garlic smells,
An' the sunshine an' the palm-trees an' the tinkly temple bells;
On the road to Mandalay,
Where the flyin'-fishes play,
An' the dawn comes up like thunder outer China 'crost the bay!"

"A Desperate Battle at Padaung"

Lieutenant-Colonel Musgrove now took over the Commando and Force Viper. On 27th March, when the Division was heavily engaged at Shwedaung on the East Bank, the Commando received orders to hold Padaung on the West Bank and to prevent the enemy from crossing the river. Two platoons and a Vickers section from Force Viper were landed. The flotilla remained inshore with skeleton crews. All was quiet on the Padaung side, although the sound of battle could be heard on the East Bank. The villagers appeared friendly, selling the party ashore bread and fowls to augment their supper. But the Burmans had concealed the Japanese in their houses, and shortly after midnight the enemy emerged.

A desperate battle ensued. Major Johnston described it thus: "At 0030 there was a burst of Tommy gun fire in the compound below us. The Colonel, Fayle and I dashed down to see what was happening. There was bright moonlight and the compound was quiet enough, except for a subdued scuffling going on outside. I went out on the road and saw figures quietly crossing towards the compound farther down. Ten yards from me a figure was kneeling on the road, and another lying on the edge of it. I had a good look at him and decided that he was a Jap. I fired a revolver shot at him, but missed, and jumped behind a gravel mound. Meanwhile, the far corner of the compound nearest the river was filling up with troops. The Colonel and I got all the reserve platoon that we could see and sprinted 50 yards down the road to where it crossed a dry gully. As we went, there was a yell from the Japs and Tommy guns, automatic weapons and rifles opened fire. I had lost touch with Fayle. We took up positions along the gully and fired back into the compound, which was now seething with troops. After a little while we were fired on from behind as well as in front, and the air was thick with the crack and thump of bullets and gunfire. Our ammunition began to run out. One Bren had none left, another had only one magazine. There was no hope of contacting Cave's platoon in the mélée and the Colonel decided that we should fall back and at least get well out of harm's way. Empty weapons and blatant heroics are very foolish confederates. We made our way back delicately, half expecting to meet parties of Japs, but there was no further excitement and we reached the ships by 0430. By 0830 the whole of Number 2 Platoon had returned – they had not had a single casualty."

Lieutenant D. R. Fayle, who was in command of his platoon, was the last to return. During the attack he had found himself with a Corporal and one Marine. They opened fire on the Japanese at short range. The Japanese tried to rush them, but without success. Then a Japanese officer called upon them to lay down their arms. He was immediately shot. Again the Japanese tried to overwhelm the small party, who continued to fire as targets presented themselves until they ran short of ammunition, when they escaped unnoticed to the river bank, embarked in a canoe, and began to paddle upstream. Then, of course, in true-blue Hollywood "Laurel and Hardy" fashion, the canoe sank under them. Murphy's Law, if it can go wrong it will, and it did. They regained the bank amid much hilarity and made their way back to the flotilla on foot, out of ammo, laughing uncontrollably, wet through and bedraggled, but alive. Well, I guess it also happens in the Army, but in the Marines when the chips are down and the wolf is at the door, it never ceases to amaze me how soldiers seem to search out the funniest parts of the most dangerous and God-awful situations and use it to their advantage. It must be that old saying again: 'You've never lived until you've almost died'. Amen to the military sense of humour.

In all, the Force lost 35 of its number that night, including Lieutenant P. Cavc and thc Vickers section. With the exception of the Delta Guard 9, however, the flotilla was kept intact by means of reduced complements.

On 4th April, at Allanmyo, the Force was attached to Burdiv, which had taken over from the Seventeenth Division. It consisted of the 1st and 13th Brigades on the east side of the river and the 2nd Brigade on the west. The flotilla continued its work, patrolling the river while Burdiv withdrew to the north, destroying boats and cutting teak rafts adrift, fighting small but important actions with the enemy on the banks, blowing up petrol barges, making contact with lost troops and, later, picking up refugees, and the staff and patients from a casualty clearing station. Sometimes the vessels were fired on from the shore, sometimes machine-gunned from the air. The Japanese broadcast a threat that if any of the Marines were captured, they would be roasted and cut into small pieces. That was the measure of Force Viper's success.

Some of the vessels had to be replaced, among them the Stella, whose engines could do no more. Her Chittagonian engineman burst into tears at seeing his ship destroyed. The time came when most of the Chittagonians had to be given their chance to return to India. Marines took their places. This meant a further reduction in the fire-power of the launches, but a gunboat, the small paddle-steamer Viking, was sent down from Mandalay to increase the flotilla's strength.

Force Viper in Burma. It is 7th March, 1942. The Japanese are advancing towards Rangoon. Three miles south of the city the oil tanks of Syriam have been fired. Billowing smoke is reflected in the Rangoon River, where the launches of Force Viper, one of them in the foreground, cover the demplition parties.

(1) The 106 men of Force Viper were marines ; they had volunteered in Colombo for special service in Burma. Two of them, bearded, are seen above in the Doris with the Chittagonian captain of the native crew. These crews tended the engines of the launches.

(2) Above is the improvised flotilla of touring launches and motor boats from which Force Viper fought. It is now mid-March. The flotilla has passed into the Irrawaddy, where (below) the oil stores of Yenanyaung, Burmese oil centre, are being destroyed.

(3) Now the flotilla sails farther upstream, protecting the army from Japanese infiltration by river. The land fighting was going badly, and turning into the Chindwin, Force Viper (below) covered the crossing of the British land forces from the east bank to the west.

(4) Their ten weeks' voyage of more than 600 miles from Rangoon over, the marines set out on 13th May from Sittaung to India. The last party left Tonhe, still farther upstream, on 15th May. Ten days later, 48 of the original party gathered in Calcutta.

By 25th April, the military situation had become even worse. A Japanese force was reported to be only a few miles from Lashio. It was imperative for Burdiv to cross the river at Sameikkon, and the flotilla ferried it over. The chief problem was the transport, which consisted almost entirely of bullock carts, but by towing a large "flat" continuously to and fro across the river from the afternoon of the 27th until the morning of the 29th, 320 carts, many of them loaded with stores, 640 bullocks and about 500 mules were ferried, while other vessels of the flotilla were taking troops and equipment across. The Viking, anchored in midstream, acted as an anti-aircraft guardship.

"Up the Chindwin and to India"

The course of the troops was now up the tributaries of the awe-inspiring Chindwin River, which Force Viper was ordered to patrol. The water was too low at this point in time for the flotilla to enter, and all the vessels, with the exception of the small motor-boats, had to be destroyed. Sternwheelers were employed as far as Shwegyin. A troop of anti-tank guns was attached to the Force, which was ordered to dominate the river and cover the troops during the crossing of the Chindwin at Kalewa on their road to India.

The troops were ferried across by the sternwheelers, one of which was manned entirely by Marines; the others had Marine detachments to act as guards, anti-aircraft sentries, and stand-by engine-room crews. By the night of 10th May the whole Division had been ferried, including the Armoured and the 48th Brigades, which were acting as the rearguard. As the flotilla got under way from Kalewa, the Alguada broke down and had to be destroyed. She had been the first vessel the Force had acquired and was almost the last it was to use, and she had done more running than any other boat in the flotilla.

The last mission entrusted to the Force was at Sittaung on 14th May, when one motor-boat was ordered to take a lakh of silver rupees and a secret letter to the District Commissioner at Homalin, 86 miles up the river. On the way upstream, Major Johnston learnt that the Commissioner had already left Homalin, but he decided to carry on and bring back refugees, who were already streaming down the river – in boats, on rafts, and on foot upon the banks. Many of them were in a dilapidated and pitiable condition. They had nothing but the rags they stood up in, but then, some could not even stand up. Many had been robbed by dacoits. Others had dropped down dead on their long journey to the unknown.

Major Johnston found Homalin almost deserted save for looters. He returned downstream, giving out handfuls of rupees to refugees on the bank – much to their astonishment. He was obviously one of those rare men who lived by the old adage: "Pass me the wine and the dice, and perish the thought of tomorrow", to him human life being more precious than silver. He then jettisoned the old money bags and took a number of refugees on board. Fuel and oil were running low, and to cap it all the engines were in distress, Murphy's Law again, wouldn't you just know it! But by going half speed and assisted by the strong current, he reached Tonhe that evening.

At dawn next day he and his small party destroyed the motor-boat and set out on the 200-mile 'yomp' overland, crossing the Indian frontier on the second day, camping one night beside a stream of clear water glinting in the very last rays of sunlight to soak sore, aching feet in the cold life-giving liquid, with a pint of hot, sandy, morale-boosting weak tea. Who the hell needs the Savoy anyway! The next day by first light it was on and up over mountains, spending another night in a Naga village high up on the wind-swept ridge, mercifully too high for the aggravating mosquito, until at length they reached Dimapur, thence travelling in luxury on a train floor to Calcutta, where the main body of the Force, under Captain Alexander, had arrived on the previous day, 25th May.

Thus ended as strange and as gallant an amphibious expedition as any of the Corps had ever been called upon to perform. The losses had been serious. Of the original Force of 107 which left Colombo, only 48 returned. Some who had become casualties and gone to hospital, reappeared later. One at least found his way back alone and by devious courses. This is a true story on record, but after searching high and low I have not managed to dig out the finer details of what must have been a "Great Escape" in itself with the adventure patterns of another of those Steve McQueen-type characters who went for absolute success at all cost. When he reported at Plymouth, he explained that his officer had told him to make his way home if he ever became lost or detached from Force Viper. The Adjutant, being a worldly man of infinite patience, must have known what the man would answer, but he asked him anyway. "What do _you_ think your officer _meant_ by home?"

"Stonehouse Barracks, sir," he said.

"Ship me somewheres east of Suez, where
the best is like the worst,
Where there aren't no Ten Commandements
an' a man can raise a thirst;
For the temple-bells are callin', an' it's there
that I would be –
By the old Moulmein Pagoda, looking lazy
at the sea;
On the road to Mandalay,
Where the old Flotilla lay,
With our sick beneath the awnings when we
went to Mandalay!
On the road to Mandalay,
Where the flyin'-fishes play,
An' the dawn comes up like thunder outer
China 'crost the Bay!

CHAPTER 10

"THE OPERATION MUST SUCCEED"

MADAGASCAR

After the Japanese occupation of Malaya and Burma, the danger to the convoy route around the Cape to the Middle East and India made it imperative for British forces to seize the Vichy-held island of Madagascar. The combined expedition was commanded by Rear-Admiral (later Vice-Admiral) Sir E. N. Syfret, flying his flag in H.M.S. Ramillies, and by Major-General R. G. Sturges, R.M., G.O.C. Land Forces. The orders issued for the assault contained the significant words: "Withdrawal is out of the question. The operation must succeed whatever the cost."

On 5th May 1942, the first landings were made in the extreme north of the island with the object of capturing the harbour of Diego Suarez and the adjacent port of Antsirane. Thanks to the co-operation of naval aircraft from the Indomitable and the Illustrious, Diego Suarez north fell without formidable resistance, but 36 hours after the operation had begun the Land Force was held up at the approaches to Antsirane by a strongly-defended line (the existence of which was unknown to the Force Commander before the assault) stretching across the neck of the peninsula on which the town stands.

On the morning of the 6th, General Sturges, having given verbal orders for the night assault on Antsirane by an infantry brigade, returned to the flagship to ask the Commander-in-Chief for assistance in the attack, by the promise of a diversion "to take the Frenchmen's eye off the ball". As a result, the destroyer Anthony (Lieutenant-Commander J. M. Hodges, R.N.) was called alongside

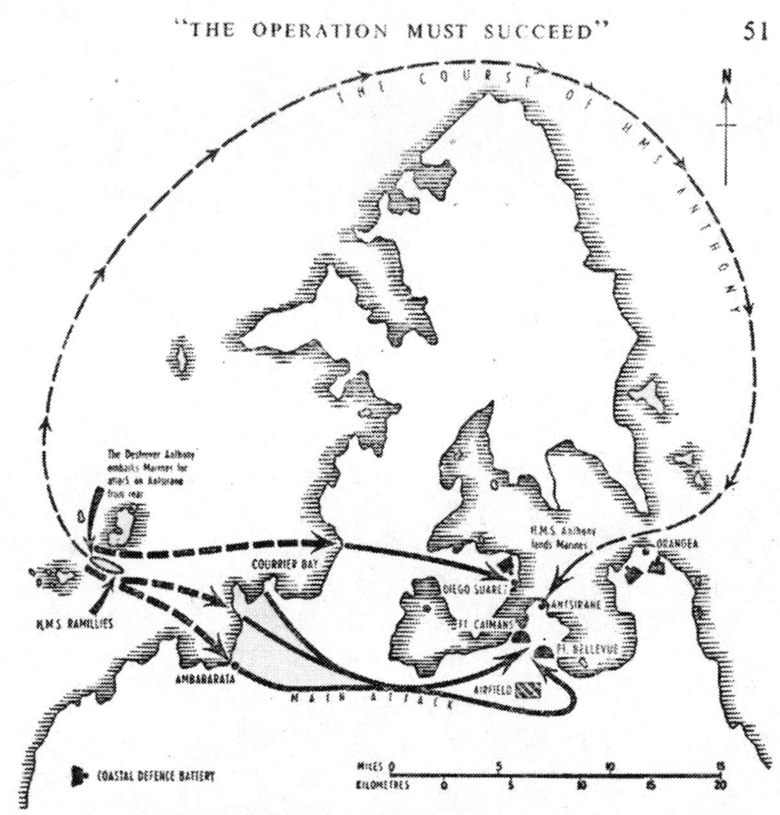

THE CAPTURE OF ANTSIRANE.

and at 2.30 pm Captain Martin Price, who commanded the flagship's Royal Marine detachment, received a sudden summons to Admiral Syfret's cabin. There he found General Sturges, who asked him how soon he could have ready a landing party of 50 Marines. Captain Price replied that his emergency platoon could be ready in 20 minutes, but that since there was no immediate organisation for a large landing party, it would take an hour. Both the Admiral and the General said it must be done in less, if the party was to arrive in time to synchronise with the night assault.

"Every Frenchman who can hold a rifle appears to be defending the neck of land on the Antsirane Peninsula," General Sturges explained. "We want you to cause a diversion by attacking the town in reverse. You will go ashore from the Anthony. Your objective is the Artillery Commandant's house, which dominates the southern end of the town. We want you to turn to the north as many rifles as possible that are now pointing south at the Army, and prevent enemy reinforcements being available for a counter-attack."

Captain Price then asked for 45 minutes. General Sturges, knowing the ability to react quickly was the Royal Marines' template, agreed and said he would postpone the night assault for half an hour.

Hastening to the quarterdeck, Captain Price ordered the Marine Bugler to sound Portsmouth Divisional call, followed by the "Fall In". The Divisional call was a profound moment for Marines afloat, since it was sounded only prior to the despatch of a landing party. Within the ridiculously inappropriate living space of a mess-deck, and believe me I know mess-decks, especially on carriers, it is a wonder to behold how every man knows the location of every piece of his personal kit from a cumbersome Bren-gun with its steel boxed magazines, right down to the safety pin for his first field dressing. From dozens of relaxed bodies lying around in what would seem like disorganised chaos, the Divisional Call would galvanise instant positive reaction accompanied by a flurry of kit shuffling activity which had become so second-nature as to prevent any one individual from hampering the progress of another, dare I say ants could probably learn lessons from 50 Marines answering a call to "Clear Lower Decks".

Following the Divisional Call the order was piped "Clear lower decks of Royal Marines. Fall in on the quarterdeck." The strength of the detachment was 160, who manned X Turret and the starboard 6-inch battery. Since the Commander-in-Chief might require the main armament to be manned, Captain Price ordered: "Starboard battery to fall in before all," as the men came tumbling aft. This gave him the number required.

Briefly he told them the task. The rig would be fighting order, steel helmets, long khaki trousers (as a protection against mosquitoes) and tropical shirts. The remainder of the detachment would collect the necessary stores and ammunition.

"The Dash in the Anthony"

Within 35 minutes of the time Captain Price had left the Admiral's cabin the Marine landing party was ready, and embarked in H.M.S. Anthony, which by that time was alongside. One sergeant, although Second Captain of X Turret, managed to become involved in the landing party. The Admiral's orderly had obtained special permission to go. Lieutenant H. G. Powell was Captain Price's second-in-command.

The Ramillies was oiling off Courrier Bay, on the west coast of the island, but the landing was to be made on the east side, which entailed a passage of 120 miles. The Anthony proceeded at high speed in a choppy sea. At 7.45 pm she began to approach the harbour and went to action stations. The harbour was known to be heavily defended and there was every possibility of encountering a minefield at the entrance. The landing party fell in on the forecastle messdeck, one platoon on either side, facing aft, ready to dash out into the waist of the ship when she was past the worst of the fire from the shore batteries.

By this time it was very dark; the moon was not due to rise until after the night assault. The entrance to the harbour was a break in the cliffs, that was all, no more than half a mile wide, and the high ground behind prevented the opening being silhouetted against the sky. Lieutenant-Commander Hodges had never seen the coast before. He had to land the Marines on time and no-one on board but he knew the odds against him. He felt as though he were coming to closure with an event, and being completely enveloped in a sense of responsibility as to how and where he put these men ashore, and therefore the lives of at least some of them after that first possible enemy contact or fire-fight at the landing polnt. But it was just his youth, time and experience would take care of that, for he knew Commanders of much larger ships probably felt just as bad when they had to put entire units ashore on enemy coastlines. Nevertheless, he found the gap and the calmer water told the landing party they had passed through. Then the shore batteries opened. The Anthony replied with her pom-poms and 4.7s. Crouching in the waist of the ship, the Marines could see little, but they could hear much of the initial contact. At one moment it seemed as though a battery were firing within a few feet of them. It induces semi-conscious reaction. It draws heads down tortoise-fashion. It sends hands to travel from dog-tags to field and shell dressing, from checking magazines on weapons, to magazines in pouches. It even makes the odd few wonder where the hell they went wrong since they left school. But above all else it prepares the senses for the inevitable conflict, like the boxer in the ring who knows the night has arrived and he has nowhere left to run. The gunfire sounded again. "You needn't worry about that," a stoker Petty Officer told them. "That's ours." It was certainly very loud and very near and those who believed him were reassured by those comfortable words – not knowing that they were quite untrue. The P.O. had only said it to put their minds at ease for he knew he was safe enough on board but these were the guys who had to hit the landing site. If there is another creature on this earth more worldly-wise than a 120 year old leather-backed turtle, it would have to be a 22 year committed R.N. stoker. They are that good.

For eight miles across the harbour the Anthony had to run the gauntlet of the shore batteries. Then she approached the jetty. The original plan had been for her to go alongside "port side to", but there was no-one ashore to take her wires, and in the darkness she overshot the mark. Wouldn't you just know it! Word was passed that she would approach starboard side to, but the off-shore wind made this impracticable also. Murphy's Law again, not to worry. By a feat of magnificent seamanship, Lieutenant-Commander Hodges then went slow astern into the jetty. The landing party disembarked from the stern, scrambling over the depth-charges. A couple of light automatics and a French .75 mm gun helped to drown the shouts of anger as Marines' shore-service boots met seamen's toes in the scuffle.

During the disembarkation, heavy machine-gun fire was coming from the jetty itself and from the hill behind. Figures were seen moving in the darkness and later the landing party stumbled over the cooking-pots of a section of native troops. While Captain Price was making a quick reconnaissance, Lieutenant Powell superintended the disembarkation and put out a covering force at either end of the jetty.

Once all were ashore, the party set off south through the dockyard. As they moved off, the Anthony steamed out of the harbour again, letting rip from her pom-pom with a full belt of tracer six feet above their heads.

"As we left the wharf," wrote Captain Price, "Fire was opened upon us, but they were obviously firing at the patter of many feet, which seemed to have grown several sizes larger than usual and to be scooping up all the tin cans and pebbles in the neighbourhood. As soon as we were clear, the shooting stopped."

"All Objectives Taken"

The dockyard was still blazing from the bombing attacks of the Fleet Air Arm earlier in the day. The Marines' progress was slow, since the column had to be dispersed to avoid casualties. At length they were confronted by a steep cliff, 100 feet high. There was no path, but they scrambled to the top, only to discover a brick wall, topped by a close-meshed wire fence, blocking the line of advance. Twenty men soon tore a gap. They then found themselves in a large compound full of workshop stores and a collection of goats, pigs and cows. They made their way to an imposing gateway, above which a tricolour was flying. A brass plate announced that this was the entrance to the "Direction d'Artillerie". Captain Price decided to make this his base for operations, while Lieutenant Powell took his platoon on to form another strongpoint from which he could create a diversion.

The sentry posted at the gate of another building (which turned out to be the Naval Depot), soon challenged the platoon's advance and rifle-fire opened. The Marines retaliated with hand-grenades. In a few minutes the inmates came out to surrender, headed by the

Commandant of the barracks and a white flag. When Lieutenant Powell had accepted the surrender, the French trumpeter with the Commandant sounded a call. He was immediately overwhelmed by Marines. The Commandant explained that he was only sounding "Cease Fire". Apologies were made and accepted.

Looking quickly through the depot, Lieutenant Powell discovered a number of British prisoners of war; three Army officers, 50 other ranks, and the crew of a Swordfish from the Illustrious which had crash-landed on the previous day. He released the prisoners, together with a British agent who was waiting to be shot next day, and armed them with French weapons, thus doubling the strength of the landing party and raising their confidence in what the morning might bring to light. In the meantime, the party outside Artillery Headquarters heard telephone bells ringing furiously within. A search revealed a telephone exchange and a number of French officers, who surrendered without a struggle, feeling, no doubt, that they had done their full duty in warning all positions of the danger in the rear. It was unquestionably this quick flashing of the news which caused the general collapse of the defence, thereby preventing street fighting in the town, or "collateral damage" I believe is the politically-correct wording in this brave new century, but I guess it did prevent loss of life on both sides. For a time, however, rifle and machine-gun fire was heard on the right flank. Captain Price took this to come from the advanced British troops and sent a section to make a contact. Actually, the fire, which soon died down, came from the enemy. The Royal Marines landing party was alone behind the French lines. It had carried out its orders and had turned the French rifles from the southward to the north.

"A Heavy Quiet that Commands"

The Land Force started to come through two hours later, and by dawn all resistance was at an end. For a time, there was a strange silence as situations were assessed, thoughts were gathered and then prisoners were pouring into the Marines' hands in embarrassing numbers. They eventually managed to palm off 500 of these on the Army "in exchange for something to eat". The battle for Antsirane was now over. Treaties and agreements were drawn up. H.M.S. Ramillies came steaming into the harbour. But as yet there was no rest for the Royal Marines. One platoon occupied the dockyard to prevent sabotage; the other turned out a guard to a regiment of French artillery which had surrendered with the honours of war. "The atmosphere after the battle was strange," wrote Captain Price. "It was as though we had won a hard game of rugger, and neither side appeared to have any ill-feeling." Night leave was granted to all (French) prisoners of war.

It was not until 8 am on the morning of the 7th that the last of the landing party returned to the Ramillies, where they received a warm welcome from the whole ship's crew. Even then they found that the easy day to which they had been looking forward was not to be, for they were called upon to form a guard of honour for the dignitaries who were to sign the formal treaty of capitulation at the Residency at 10 o'clock. At 9 am, fed, shaved, clean, very different from the unkempt and weary body of men who had crawled over the side an hour before, they set off ashore, and headed by their band, marched to the position for the guard.

The conference started at 10 o'clock and ended at 2. The sun was very hot and the guard was very tired, having had little or no rest for two nights. But as they returned to the landing stage through the stifling town, with the band echoing and crashing in front, they marched as proudly as if they had been on parade at Eastney Barracks. Some of the watching troops, who were seeing the Corps on parade for the first time, looked in astonishment at their white cap-covers, their red cap-bands, their clean tropical rig, and at the officers with their archaic swords, and wondered why these old-fashioned troops were marching in ceremonial fours. But there were some who knew why General Sturges had chosen them for that honour. One of these was a long-service private soldier, who stood to attention by the roadside as the guard marched past, and was heard to repeat at intervals: "Good lads, good old Marines, good lads!"

It was the praise that Rudyard Kipling rated so highly; the praise of the fellow professional. Rear-Admiral Syfret, in an Order of the Day, described the task entrusted to the Royal Marines as one "requiring supreme self-discipline, courage and unselfish devotion to duty", and declared that they had performed it with complete success. And General Sturges, in a private letter to the Adjutant-General, Royal Marines, mentioning that he thought the chances when they left the ship were "four to one against", said: "Make no mistake, we were proud of them!" Captain Price received the D.S.O., Lieutenant Powell the D.S.C., one sergeant and two Marines the D.S.M. Great credit for the success of the operation must go to Lieutenant-Commander Hodges, who was also awarded the D.S.O. "The Anthony carried out her duty with a stout heart and splendid efficiency," declared Admiral Syfret, and added: "The work of the Anthony and the landing party was worthy of the best traditions of the Royal Navy and the Royal Marines." It was, indeed, a perfect example of the manner in which the two work together, which is the reason why I chose it from a selection of other gallant landings made by detachments from H.M. ships during the subsequent course of the Madagascar campaign.

CHAPTER 11

IN DEFENCE OF THE MERCHANT SHIPS

"SELF-DEFENCE - NATURE'S OLDEST LAW"

Besides co-operating with land forces, as in Malaya, Burma, and Madagascar, the Royal Marines have had many opportunities of giving direct protection to the ships of the Merchant Navy, in addition to manning the guns of the naval forces which cover the passages of important convoys.

For a few weeks after the fall of France, three vessels flying the White Ensign, but manned almost entirely by Marines, accompanied the Channel convoys as additional escorts against the E-boats which were becoming an increasing menace in coastal waters. These vessels were Dutch schuytes – fishing boats of about 200 tons – which had escaped from Holland during the occupation. The Admiralty took them over from the Netherlands Government, armed them with pom-poms, 6 pounders, and Hotchkiss guns, and formed them into a flotilla under Commander Sir Geoffrey Gongreve, Bt., D.S.O., R.N. The second and third vessels of the flotilla were commanded by Lieutenant D.B. Drysdale and Lieutenant G.J. Bower, who were the first Royal Marine officers to be in charge of ships flying the White Ensign during the Second World War. Each schuyte had a skipper, R.N.R., as navigator, and 20 Marines to man the armament; also two helmsmen and two enginemen, who were naval ratings.

The flotilla began operations in September 1940, and patrolled in the Channel every night. The Battle of Britain was still in progress, and on more than one occasion the schuytes had to "shape a course between two air raids", as one of the officers put it. Their shallow draught made them immune from torpedo attack (or so it was believed) and they would "button on" to the convoys with the object of attacking any E-boats which came within range. The E-boats kept their distance, however, and thus the flotilla, by accompanying the convoys through the most dangerous areas, gave the merchant ships confidence, and also drove off a number of enemy air attacks. These operations lasted six weeks, after which the flotilla was disbanded, the vessels were put to another use, and the gunners employed elsewhere.

"Back to the Colours to Fight the Guns"

Other Royal Marine gunners had been helping to defend merchant ships, whether in convoy or sailing independently, since the beginning of the war. In the early days most of them were pensioners with previous war experience. Some were over 50 years of age. Together with the Royal Fleet Reservists and other naval ratings, they made up the organisation known as D.E.M.S. (Defensively Equipped Merchant Ships). This was subsequently extended to include Army gunners, who then formed what became known as the Maritime Regiments of the Royal Artillery.

The first of these Marines and naval pensioners to serve in merchant vessels were the only trained gunnery ratings on board and consequently they had to assist in training members of the crew to serve and fire the guns, which ranged from 12 pounders to 6-inch, machine-guns being added as they became available. The guns' crews were composed of officers and upper-deck hands, with volunteers from the engine-room. In ships trading to the East, some of the Chinese or Lascars were used as ammunition numbers.

The pensioners signed on as deck hands. They fought the guns of every class of ship in the Merchant Navy and of many which flew the flags of the Allied Nations. They fought in the Atlantic, in the convoys to Russia and Malta, on passages to and from the Far East, on the Cape route to the Middle East and in the Indian and Pacific Oceans. Many of them saved their ships against the attacks of Axis aircraft and submarines.

Marine H. Calcott was awarded the B.E.M. for his devotion to duty when one of the Elder Dempster Shipping Company's vessels was chased by a U-boat. She was zig-zagging at about 10 knots in clear weather when a large enemy submarine was seen to be pursuing her on the surface at high speed. At about 10,000 yards the U-boat opened fire without a warning shot. The merchant ship used smoke floats, which seemed to hamper the enemy, so that none of the shells reached its mark. When the U-boat was within 7,000 yards range the steamer hoisted her Ensign and Marine Calcott opened fire with the 4-inch gun. The third shot detonated on the U-boat. She dived immediately and was not seen again.

Throughout my investigations for this book, I have to say this was just one of very many instances of submarines being compelled to abandon a chase through the steadfastness of the D.E.M.S. gunners and their crews. For another example, while a vessel of the United Merchants Steamship Company was towing to port another which had been put out of action by enemy aircraft, the ships were attacked by a Dornier 17 with bombs and machine-gun fire, but the gunners drove it off. The Master of the towing ship, in a letter to Commandant of the Plymouth Division, paid a tribute to "the sterling ability shown by Marine Chedgey, who not only assisted the Chief Officer in our efforts to connect to the damaged ship, but displayed that cool, efficient stability and marksmanship which made the enemy's attack abortive. His steadiness and his control of the guns' crew were a credit to the Corps and but for this gallantry, I am afraid all our efforts to save the vessel at the eleventh hour would have failed".

Time after time the D.E.M.S. gunners and the crews they had trained had defended their ships by standing to their guns with bombs falling around them, never missing the chance of a shot. Often they had damaged and driven off the aircraft, but there had been occasions when the number had been too many for them and they have had their ships sunk under them, firing their guns to the last.

"With the Channel Convoys"

Such was the fate of S.S. Terlings, a vessel of 5,000 tons belonging to Messrs. Lambert Brothers. She was sailing in a coastal convoy of some 30 vessels, with only a trawler escort. Her gunlayer was Marine W. C. Prescott, who had been attendant to King George VI when His Majesty was a midshipman in H.M.S. Cumberland. The crew of her 4-inch gun consisted of English deckhands and Lascar stockers.

As the convoy passed through the Straits of Dover, it was shelled by the German batteries on the French coast, without casualties. Off Portland a formation of 38 German dive-bombers came in to attack. They sank two small oilers, then made a concerted attack on the Terlings, dropping over 100 bombs. There were nine direct hits on the engine-room, and three on the bridge. All the engine-room officers were killed, and most of the Lascar stokers; also the cook and two cabin boys. Many others were blown into the sea. Marine Prescott, who was subsequently awarded Lloyd's War Medal for Bravery at Sea, thus described his own part in the raid:

"My gun's crew were all machine-gunned by low-flying aircraft. Mr. Smith, the Third Officer, had a machine-gun bullet through his neck, but carried on as long as he could. Mr. Ludlow, the Chief Officer (later B.E.M. and L.M.), came aft to help man the gun. The bombs had blown the boilers and bottom out of the ship. She started to sink, but the after part of the deck, where the gun was mounted, was still above water.

"The Lascars behaved marvellously. You treat them fair and they'll treat you fair and 'play their part'. Two of them were killed at the gun, others blown off the gun-platform into the sea. I clung to the gun. The Third Officer was at the training wheel and he clung to that, even after he had been wounded. We were shrouded in black smoke and gagged on the stench of oil and explosive residue. We fired two more rounds after that. The Chief Officer up-loaded. Then the magazine went under water. Our ammunition supply was now gone and we had no crew. The Captain shouted from a hatchboard in the sea: 'Come on, guns, you can do no more bloody good there.' I then dived into the ditch with the Chief and Third Officer, and got hold of a board. She sank as we left her.".

"While we were in the water the German aircraft came back and machine-gunned us. They shot the Captain in the back as he lay on the hatchboard. I got a bullet in the thigh. Everyone was wounded somewhere. I was in the water about three hours and was picked up by the Destroyer Scimitar."

As I mentioned earlier, I selected these tales from many such actions and the names mentioned are but a fair representation of those Marines and Corporals, Sergeants and Colour-Sergeants, all of them men in middle life, who returned to sea to place their professional skill at the service of the Merchant Navy. A large number of this old guard was later withdrawn from sea service to train the younger generation of naval ratings who began swelling the ranks of D.E.M.S.

"Winnie and Pooh"

Convoys sailing in home waters have also cause to be grateful to the batteries along the coast. At no point has their support been more effective than that given by the great naval guns, which were mounted near Dover and manned by Royal Marines, as an answer to the long-range batteries across the Straits.

After the occupation of the French coast, these German batteries threatened to interfere with "Churchill's Armada", as the convoys sailing through the Straits had come to be called, owing to the Prime Minister's insistence that they must continue to sail the English Channel. To give them protection against the German guns, the Royal Marines Siege Regiment, at the direct insistence of Mr. Churchill, began operations at the end of June 1940. The first gun, a battleship type, was installed in six weeks, and was appropriately called Winnie after the Prime Minister. The second, named Pooh by an obvious association of ideas, followed some weeks later.

These gigantic guns, which had a "super-super" charge of cordite, never fired less than 20 miles. At first they were used for counter-battery work when the convoys were passing through the Straits; later they were used against coastal targets. They were also held in readiness for defensive purposes in the event of invasion. They had been both shelled and bombed, but without damage or serious casualties.

Winnie and Pooh were static guns, but some distance from them were heavy naval guns on railway mountings called by such names as Piecemaker and Sceneshifter; their range was approximately the same as that of Winnie and Pooh. They were used mainly for cross-Channel counter-battery work, but they could also cover targets along the English coast.

When the Siege Regiment was first installed, the men were mainly Continuous Service Marines. These were gradually withdrawn for sea service as new gunners were trained, and later all the officers, and the great majority of the men, were drawn from those who were serving for hostilities only.

Royal Marines of the same category would also man the forts in the Thames Estuary and elsewhere, which protected the East Coast shipping from mine laying aircraft. These forts were commissioned as H.M. ships and were called after the sands on which they stood. They were constructed so that they could be towed out to the required position, whereon the base was submerged. Each fort consisted of two concrete towers, 50 feet high from the base, connected by a steel superstructure on which the anti-aircraft guns and equipment were mounted. Fixed to the base was a landing-stage, made of steel joists. The mess decks were in the towers, with a store-room and magazine below, the ammunition being brought up by lift. The forts were connected by telephone to the shore.

All the armament was manned by Royal Marines under R.N.V.R. officers, with naval ratings for technical duties. The guns and equipment were supplied by the Army, and the forts were in close contact with the R.A.F., thus forming an interesting example of inter-service co-operation. They accounted for a number of German aircraft. The Royal Marines also provided their own signallers. There is another small story I found within the dog-eared reams of naval history of one rough morning soon after the installation of one of the Thames forts, when the signallers were less proficient than they later became, a destroyer was seen entering the estuary. A long hoist of flags went up as she approached the fort. Laboriously the Marines on duty spelt out the signal. When they had finished, it read:

"And how are the Little Princes in the Tower this morning?"

By the time the Commanding Officer had received it, the destroyer was too far away for him to frame a suitable reply. But all concerned knew that this sort of interaction between Navy and Marines was long-standing and traditional, for as Admiral Sir Andrew Cunningham told his staff in Alexandria, in May 1941:

"It takes the Navy three years to build a ship, but it takes three hundred to build a tradition."

CHAPTER 12

USEFUL MEN TO HAVE ABOUT

The Royal Marines have always shown that they have an aptitude for co-operation with others. They are equally proficient, as Kipling pointed out, at paddling their own canoes; and when there are rapids ahead, they shoot them better than most.

They are, you might say, useful men to have about a ship; and they are equally useful men to have about ashore. That is why you will come across Marines all over the world doing special jobs which require initiative and self-reliance. It is a kneaded and moulded second nature to them. This is the result of their lengthy training in the gentle art of improvisation. They have an absolute genius for it. The Corps teaches them many valuable lessons, and good manners is one of the most abiding. They are trustworthy, efficient, good-humoured, and polite. Thus they are in demand for filling posts where those qualities are needed in many delicate areas, both at home and abroad. During the second World War, many of these men were pensioners. Some joined the Royal Marines Police – a force which was started after that war and was open to naval reservists and to direct entries from civil life – and were to be seen in H.M. Dockyards, guarding naval ammunition dumps, or on duty at naval air stations and other establishments which came under the White Ensign. Some were on the staffs of Naval Railway Transport Officers, when they invariably lightened the way of harassed seamen who were more at home in a ship than a railway station. One colour-sergeant, who had spent three years at a London terminus, was once asked how he liked his work. "Well sir," he said, "There's one thing I have learned from it, and that's patience."

The training of the Royal Marines makes them particularly suitable to act as orderlies for special duties abroad. At Madagascar, Major-General Sturges had a corporal and six Marines attached to him from H.M.S. Ramillies. Vice-Admiral Sir Harold Burrough had others in personal attendance when he was commanding the Inshore Force during the North African landings. When Lord Beaverbrook went to Moscow in September 1941, he took three corporals to keep guard over the confidential books. And when the Prime Minister attended the Casablanca and Washington conferences, special parties of Royal Marines accompanied him.

During the course of all the aforementioned and many more escorts and attendances, one or two which I have been advised, even now, not to talk about, there was not one instance of a failure of etiquette by the Marines.

Winston Churchill was once heard to remark of "how they do their particular line of work so much better than anyone else, and this was the margin of their success on duties well performed, and days well spent".

"Hard Months in Malta"

Marines had always been employed as guards and orderlies in naval bases, both at home and abroad. There had, for example, for many years been a detachment stationed in Malta, with its headquarters in H.M.S. St. Angelo (previously H.M.S. Egmont), one of the oldest forts on the island. In peace time the detachment provided the Admiral's orderlies, also signallers, wardens at the detention barracks, and staff for the rifle ranges. After the outbreak of war, it was given the duty of manning Lewis gun posts in the naval dockyard. Two of those in H.M.S. St. Angelo were also manned by Marines.

High-level bombing began at 7 a.m. on 11th June 1940, the day Italy declared war, and continued intermittently for several months, but the Italians flew so high that they presented no targets for the Marines' close-range weapons. It was not until January 1941, when the damaged aircraft-carrier Illustrious was in harbour, that the island had its first experience of intensive dive-bombing from the Luftwaffe. The Marines were then given two Bofers guns, which were mounted on the Upper Barraca, one of the highest points in the Grand Harbour. By collecting all the available N.C.O.'s and signallers, an officer's assistant, a boy-bugler, and one of the Admiral's orderlies, it was found possible to muster two guns' crews, and one for reliefs and replacements. The men had never fired a Bofers before, but learnt after a week's training under a Royal Artillery instructor. This self-constituted battery remained under the Senior Officer of the detachment, Captain F.F. Clark, and operated as an independent unit.

The guns fired for the first time on 4th February. Usually, it was not possible to estimate the success of an individual gun owing to the number firing from the concentrated harbour defences, but in the early hours of 28th February the battery claimed its first definite victim: a mine-layer, believed to be a JU.88, was hit, and was last seen diving steeply and clearing the breakwater by only a few feet.

Throughout 1941 and until May 1942, the raids continued with varying intensity. April 1942 was the worst month, when there were three raids a day with unfailing regularity: at breakfast, at rum-time (the hour had been altered) and at 5.30 pm, delivered by 60 bombers or more, escorted by swarms of fighters. These aircraft were nearly always German; when the Italians took part the German fighters came behind, as though driving their allies into action. The R.A.F. fighters were completely outnumbered, and in spite of their valiant efforts against overwhelming odds the main weight of the enemy's attack fell on the anti-aircraft guns. "The gunners had to sit down and take it, as well as hand it out," wrote Captain Clark. Now most of the men were beginning to understand the infamous words of Captain Scott of the Antarctic when he said, "God, this is a terrible place."

The last heavy day raid was on 10th May, by which time there was an acute shortage of ammunition, so that instead of firing 400 rounds a day, the Royal Marine battery could use only 15, firing with one gun which, nevertheless, shot down a JU.88 into Dockyard Creek. Stone cold dead in the water. But on that day the island had air superiority for the first time: 80 Spitfires, they came in as welcome as the spring clatter of re-birth after a long, cold winter of wanting, and destroyed 16 enemy aircraft. A fine taste of Murphy's Law for the Luftwaffe. One of these, a Cant, broke into so many pieces that the Marines described it as a leaflet raid. "It was worth a couple of guineas a minute to watch," they said. This was the turning point of Malta's long battle with the Luftwaffe and was the last occasion on which the Royal Marine anti-aircraft battery was in action. It can hardly have had a more successful climax.

With the coming of the Spitfires and the arrival of more ammunition, the raids decreased in violence, and on 20th May the Marines turned over their Bofers guns to the Army. A ceremonial parade was held, when Vice-Admiral Sir Ralph Leatham addressed them, and the guns' crews, who had provided the guard for the occasion, marched away to the strains of "A Life on the Ocean Wave", played by a band of the Royal Malta Artillery.

During those hard months of conflict, the Royal Marines had had surprisingly few casualties, although their sergeants' mess was twice demolished and the fort received over 60 hits from bombs of all calibres. Near-misses in the adjacent creeks were too numerous to record, even in one day's activities; on one occasion a Bofers gun-pit, 120 feet above sea level, was put out of action by the mud and water from a bomb which fell in the harbour.

In all, or perhaps I should say from all the records I could find on these actions, the Royal Marine battery destroyed or damaged 50 Axis aircraft. Captain Clark was awarded the D.S.C. and Bar; his battery won four D.S.M.'s and five mentions in despatches.

In Defence of Malta. Through many of the island's darkest months, from January 1941 to May 1942, marines manned Bofors anti-aircraft guns above the Grand Harbour at Valletta. The picture shows the ceremonial parade on 20th May when, their duties completed, the marines turned their guns over to the Royal Artillery.

CHAPTER 13

FLYING FOR THE FLEET

During the Battle of Malta, Royal Marine pilots in the Swordfish and Albacore squadrons based on the island played a notable part in attacking axis shipping in the Mediterranean. It was no innovation for them to fly with the Fleet Air Arm, for the Corps had provided the Fleet with a number of pilots since the inception of naval flying.

Lieutenant E.L. Gerrard, R.M.L.I., was one of the first four officers to be trained as a naval pilot at the Royal Aero Club's airfield at Eastchurch in 1911. When the Royal Flying Corps was established in the following year, more Royal Marine officers were trained for duty with the Naval Wing, and on the outbreak of war in 1914, by which time the Royal Naval Air Service was in being, 12 had qualified as pilots. Many more joined during the war; others served as observers, and Royal Marine N.C.O.'s assisted in training recruits in the elements of drill and discipline. Lieutenant Gerrard won the D.S.O., and Captain R. Gordon, R.M.L.I., the second officer to qualify, distinguished himself in the destruction of the Königsberg in East Africa and at the Battle of Ctesiphon in Mesopotamia. He was awarded the C.M.G., and both he and Lieutenant Gerrard retired as Air Commodores in the R.A.F.

In the second war, the Royal Marine pilots of the Fleet Air Arm maintained the high tradition of their predecessors. Lieutenant G.B.K. Griffiths was one of the first three pilots to bomb an enemy submarine, the U-30, which had attacked S.S. Fanad Head in the North Atlantic on 14th September 1939.

"They Came like Lambs"

During the Norwegian campaign other Royal Marine pilots from the Ark Royal performed excellent service. Captain E.D. McIver, who took part in a raid on shipping in Bergen Harbour on 14th April 1940, followed his leader in to the attack and dropped his bombs with great skill in spite of the appalling weather. He failed to return. Captain N.R.M. Skene won the D.S.C for leading a Swordfish squadron in two bombing attacks on Vaernes Airfield at Trondheim and destroyed three hangars in the face of intense anti-aircraft fire.

On 27th April Captain R.T. Partridge, flying a Skua, shot down a Heinkel III while on patrol in the Aandalsnes area and soon afterwards was himself forced to land in four feet of snow, owing to engine trouble. Having destroyed the Skua, Captain Partridge and his observer, Lieutenant R.S. Bostock, R.N., sought shelter in a hut nearby. While they were investigating inside they heard a whistle, and on looking out saw three of the crew from the Heinkel they had shot down; it transpired later that the rear gunner had been killed during the attack. They were all armed with revolvers and knives.

For a few moments the two parties stood looking at one another in silence. To the British airmen the situation appeared unpromising. They had no obvious weapons of any kind. Did they have concealed weapons they could reach for! The Germans had no way of knowing, nor did they really seem to care. They seemed to be enveloped in a belief of the old adage: 'Well, an eye for an eye will only leave us all blind', so momentarily nothing happened. At length, Captain Partridge knew it was time to take charge of this dangerous vacuum, stepped forward and said to the Germans, "You come in here." There was a slight pause. Then, "In they came like lambs – to our great relief," as Lieutenant Bostock said afterwards. Once they were inside the hut, Captain Partridge continued to take charge, making himself understood in a mixture of English and what could only be described as indifferent German. Eventually, he informed the Germans that they were to spend the night in the hut, and that he and Lieutenant Bostock would sleep in an hotel which could be seen at the bottom of the hill. The Germans might join up again in the morning, he said.

The hotel proved to be deserted. There were broken windows and warped doors that did nothing to keep out the bone-piercing wind chill, but it was precious shelter with a good supply of cut wood for an old pipe-stove. Early next morning, Partridge and Bostock were preparing breakfast from some porridge they had found in the kitchen, when the Germans re-appeared, saying, "We have come as you told us, please." They still had their weapons and even though they now knew the British were not armed, the whole war situation seemed completely irrelevant to them. They were enthralled by the smell of sugary porridge, and the two officers knew there was nothing for it but to ask them to share the breakfast.

I wonder what they would think of margarine lumps dipped in cocoa! A Royal Marine special for those in dire straits.

After breakfast, Partridge announced that he was going to explore the country in the neighbourhood of the hotel. The Heinkel's navigator offered to go with him. Bostock stayed in the hotel to keep an eye on the remaining two Germans who were now making more porridge on the well burning red-hot stove. Soon after leaving, the two explorers ran into a Norwegian ski patrol. On being challenged by the well-armed patrol, the German navigator's hand foolishly went to his revolver. A rifle snap-shot rang out and he fell dead on the snow. Later, Partridge was to remark how in the mere millisecond following this action, he was overcome by a sense of complete waste of a life because this man was stupid enough to reach for a revolver in the face of a much more powerful patrol. But he also knew to try to understand how a trained soldier will react in this sudden "red mist" of conflict would be akin to understanding why it always rains on bank holidays. It is a highly personal reaction of self-defence, which will never be controlled by any area of the ever-fluid rules of engagement, but this argument is for another day. (I pushed this point in Ireland and lived to regret it.)

On hearing the shot, the others emerged from the hotel. The Norwegians kept them covered, while Partridge tried to convince them that he and Bostock were English, without much success until Bostock, in a moment of inspiration, pointed to the tailor's label inside his monkey-jacket, saying, "Look, Gieves, London." The head of King George V on a half-crown produced by Partridge clinched the matter, and it then appeared that the English-speaking leader of the patrol had a brother-in-law serving in H.M.S. Glorious, who was a close friend of Partridge's. Proper little League of Nations!

The Germans were less fortunate. The patrol marched them off, protesting, to French headquarters. Partridge and Bostock, after an abortive attempt on skis lent by the Norwegians, made their way to

Aandalsnes, which they reached just before the evacuation. For his "initiative and resource" Captain Partirdge was awarded the D.S.O. – the first decoration to be won by the Corps in the second war. He was taken prisoner during a raid which the Ark Royal's Skuas made on the Scharnhorst in Trondheim Harbour, the last raid of the campaign.

Another Royal Marine pilot, Captain W.H.N. Martin, led a squadron attack on a railway line east of Narvik on 9th May. From all I managed to gather from the records, this is his account of the raid, which was broadcast to the ship's company of the Ark Royal:

"When we arrived at a point near Narvik we were greatly encouraged to see one of our fighter patrols overhead. We split up into two sections and proceeded independently to our pre-arranged targets. Captain Skene's section went up to a large railway bridge near the Swedish border and each aircraft carried out two attacks on this objective and on a tunnel nearby. One pilot obtained two magnificent hits right in the middle of the track on the bridge and then Captain Skene placed a salvo into the mouth of the tunnel. My section went to a place called Hunddallen, where we were greeted by the pleasing sight of a train standing in the railway station. This we attacked, and I managed to hit the train with a Z50-LB bomb. It caught fire and the front half overturned. We were greeted with flack and machine-gun fire, but although two aircraft were hit, there was no damage to personnel."

Since those early days, pilots of the Royal Marines have flown with no less gallantry, both from the decks of aircraft-carriers and from shore bases. One took part in the battle of Taranto and was awarded the D.S.C. The same officer, Captain Oliver Patch, won the D.S.O. for leading a sub-flight of Swordfish in an attack on the Italian warships in Bomba Bay, on the Libyan coast, when he himself torpedoed a submarine and the remaining two aircraft accounted for another submarine, a destroyer, and a depot ship. This "brilliantly conceived and most gallantly executed attack", as Admiral Sir Andrew Cunningham described it, was related in the official publication "Fleet Air Arm". It is enough to say here that the sinking of four warships with three torpedoes was a feat which was unlikely to be surpassed in naval air warfare.

"Albacores in the Desert"

Royal Marine pilots have also been decorated for destroying enemy bombers and shadowers during the passage of the great convoys to Malta, and from June to November 1942, Major A.C. Newson commanded an Albacore Squadron which co-operated with the R.A.F. and the Eighth Army in Egypt and the Western Desert. The Squadron's work consisted mainly of finding and illuminating front-line targets for the R.A.F. bombers, and for its own dive-bombing attacks. It was also employed for mine-laying, raids on enemy-occupied harbours, attacks on ships at sea, and spotting for the many coastal bombardments made by the Mediterranean Fleet in support of the troops ashore. Throughout this period, Major Newson and his comrades shared the difficulties and dangers of operations conducted from temporary landing grounds in the desert, advancing and retiring according to the fortunes of war.

From June until August (when the Eighth Army halted the enemy's advance at El-Alamein) the 12 Albacores of the Squadron made no less than 370 operational sorties, and their activities continued with equal intensity when General Montgomery attacked in October.

On one occasion, Major Newson was piloting one of the three aircraft engaged in spotting for a night bombardment of Mersa Matruh Harbour by two 6-inch cruisers. On his way to the target his engine started to overheat. The bombardment was due to start in half an hour, and it soon became apparent that he would either have to return to his base or jettison some of his load to gain the necessary height for spotting. Rather than go back, he decided to throw out the two long-range petrol tanks and his illuminating flares. He was thus able to gain the required height and carried out the duty assigned to him, in the face of heavy flak, with a slowly failing engine. The ensuing bombardment was successful, and for his part in the action Major Newson was awarded the D.S.C.

Later, his Squadron moved to Malta, where it shared in the task of harassing the enemy's sea-borne lines of communication from Italy to Tripoli and Tunisia. On one night alone, four of the Albacores sank or seriously damaged two merchant ships and a destroyer with three torpedoes.

At the beginning of the final phase of the operations in North Africa, the landings at Oran and Algiers, in November 1942, a specially trained fighter squadron of six Fulmars from H.M.S. Victorious, led by Captain R.C. Hay, D.S.C., performed most valuable work in Army co-operation reconnaissance. They reconnoitred the roads leading to Algiers, photographed bridges and airfields, and often made personal contact with the troops by landing near them.

Besides fighters and torpedo-bombers, Royal Marine pilots also flew the Walrus Amphibians, whose chief duty was reconnaissance and anti-submarine patrols. On one occasion, when a merchant ship had been torpedoed 100 miles off the West Coast of Africa, a Walrus piloted by Captain V.B.G. Cheesman, from H.M.S. Albatross, was on the scene in ten minutes. Having counter-attacked the submarine with depth-charges, Captain Cheesman landed near the survivors from the sunken ship. For hours he taxied to and from, encouraging the men in the water, formed them into groups for greater ease of rescue, and actually towed some of the ship's boats to within their reach. More than once he stopped his engine to jump overboard and go to the assistance of a wounded man.

Meanwhile, the sea had risen, until it was too high for the Walrus to take off again, and it had to remain with the boats. After many hours of waiting, rescue vessels arrived. When all the survivors had been embarked, the Walrus was taken in tow and after an all-night passage reached the shelter of land. The tow was cast off. Captain Cheesman started up his engine and taxied into harbour alongside his parent ship undamaged. He was awarded the M.B.E.

Thus did the Royal Marine pilots serve the fleet with their comrades afloat and ashore. Useful men to have about and aboard. Their numbers had been small, but their services had been so outstanding that the corps might well hold itself to be entitled to add two words to its motto: Per Caelum – by air.

The Landing Party Returns. These are marines come straight from battle. They have taken the defences at Antsirane, Madagascar, in the rear on the night of 6th May, 1942, and the position has fallen. The picket boat is returning them, their mission completed, to their battleship, the Ramillies.

The Last Inspection. A few hours later, these marines landed to fight in Sicily, the first sea-borne troops ashore. It was their duty to destroy coastal defences. Below, a marine makes sure that his machine-gun will be in perfect order.

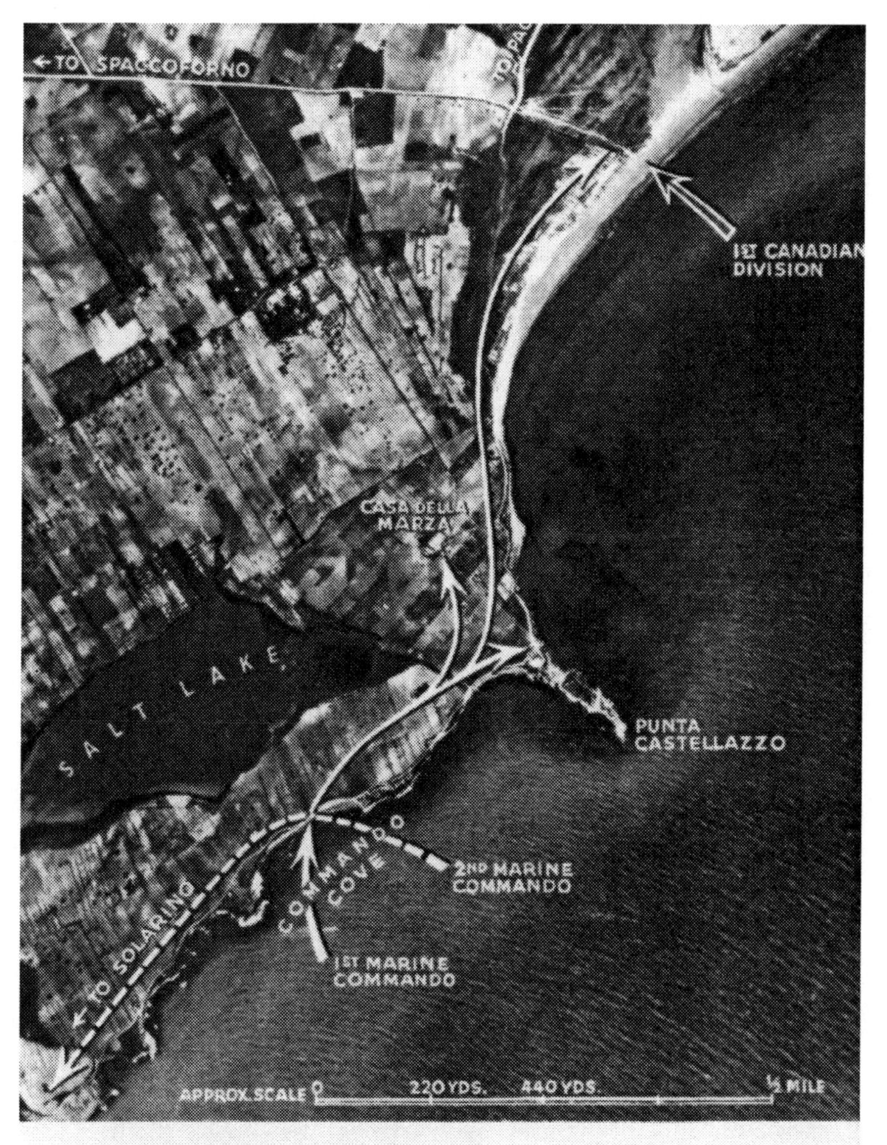

"A Sandy Bay in Southern Sicily." This air photograph shows the point of first assault in the invasion of Italy. Advancing from Commando Cove, the first Marine Commando moved east to destroy strong-points, especially Casa della Marza, that might threaten the Canadian landings. The second moved west towards Solarino. At sunrise on 10th July the area between Solarino and Punta Castellazzo was in the hands of the Commandos, and patrols were pushing inland along the road to Spaccoforno.

CHAPTER 14

THE SPECIAL SERVICE BRIGADES

Unlike most of the Marines whose activities I have tried to give a fair description of in putting this book together, the men of the Royal Marines Special Service Brigades were not normally called upon to serve afloat. They were almost entirely recruited for Hostilities Only and wore khaki, with the blue and red flash "Royal Marines" on the shoulders of their tunics or battledress.

Before we go on with our story and we are on the subject of dress, allow me to say a few words on the wearing of the beret. I must admit, on one or two occasions during my service I came close to decapitation by pacestick when a drill instructor found me with the Globe and Laurel cap badge anywhere other than directly over my left eye. When the blue beret was introduced in the Royal Marines in August 1943, it was manufactured with a scarlet semi-circular patch behind the badge and a built-in buckram stiffener. If the opening for the draw-string ribbon at the back was worn in the centre of the head, the badge was positioned half-way between the left eye and the left ear.

At this time on the green Commando beret, Royal Marines always wore their badge over the left eye. In 1949 instructions were issued in the Corps for the badge on the blue beret to be worn over the left eye also, the ribbon at the back was to be tied in a knot, and the loose ends tucked into the opening out of sight.

Blue berets manufactured after that date were made with the stiffener and patch over the left eye. The origin of the scarlet patch is also of interest to our history. Before being worn on the blue beret, it was worn on the khaki side cap in the Royal Marines from 1940.

It was a "relic" from the old Broderick cap, on which various Army regiments wore different coloured semi-circular patches. This headdress was most unpopular in the Army, lasting only about five years and abolished in 1905 in favour of the peaked forage cap. However, it continued to be worn in the Royal Marines until 1923 and consequently to many people became almost synonymous with the R.M. The scarlet patch was even worn on the forage cap until 1933, when a new cap with red band was introduced in the Corps.

Now, on the Special Service Brigades, let us go back to our story when in October 1939, it was decided to form the first Royal Marine Brigade, which was to be an amphibious unit at the disposal of the Joint Chiefs of Staff Sub-Committee. The function of the Brigade was thus appropriate to the Corps, and it differed from that of a Royal Marine landing party only in strength and in the fact that it was designed to operate from a land base instead of from a warship.

After the failure and withdrawal from Dunkirk, this Royal Marine Brigade was one of the few nearly equipped formations of its size who were capable and ready to defend the English coast. Later, it sailed on what was known as the Dakar Expedition, but did not land on French soil. It was then decided to expand this force into a division with similar functions. Two more Brigades were formed and a battalion of the Argyll and Sutherland Highlanders was attached. The third Brigade was eventually amalgamated with the first and second; certain units were sent to the Middle East and ultimately landed in Sicily with the Eighth Army; one company was detached for special duties in the Western Desert, including the ground defence of naval airfields and desert patrols with the French Foreign Legion.

"With the Functions of Commandos"

With the immediate future in mind and the growing need of forces for amphibious operations, the Royal Marine Division was then reorganised as two Special Service Brigades, with the functions of Commandos and with their own complement of landing craft entirely manned by Royal Marines. They were made available as an assault force to land on enemy territory and establish a bridgehead through which the main body of the Army would then advance, or again, these well-organised, self-contained units might deliver a series of rapier-like thrusts at the enemy's defences overseas.

These brigades, as we have established, were under the operational control of the Chief of Combined Operations, but they were paid by the Admiralty. Like all Royal Marine forces ashore, they were under the Army Act. Although their training was mainly military, the troops were not allowed to forget that they were always and ever "members", a word they frequently use even today, of the Corps whose traditions are interwoven with those of the Royal Navy. They were and are encouraged to use sea terms and are taught something of seamanship. The psychological effect upon the men was remarkable, and when they went afloat they felt that they were not strangers to the custom and usage of the sea, for, as one of their Brigadiers was heard to say, their training teaches them "not to see a lion lurking in every corner of the ship."

For would-be members, this education began as soon as they joined as new entries at the Royal Marine Training Depot, where they received six weeks' preliminary training. Then, the Depot was situated in a lovely part of Devon, and had a good parade ground, well-built huts, canteens, and a cinema. There was also a band, composed entirely of pensioners. I found on record that one recruit, writing home, described it as "a holiday camp – except for a few N.C.O.'s." But I am certain he would soon learn that at the end of a desperately hard day of training when the chips were down and there was not much left within, both physically and mentally, then it would be the N.C.O.'s who would supply the necessary glue to keep mind and body together. "Private soldiers and Field Commanders may come and go unnoticed, but where would we be without good old dependable Corporal Smith or Jones!"

These very words were spoken to me on a freezing parade ground early one winter's morning, by none other than Field Marshal Bernard Montgomery of Alamein, just before his retirement from the military in 1964. More on that later.

At the Depot some of the new entries were disappointed to find that they were unlikely to go afloat, but soon became content to serve the Corps, of which they so quickly learned to be proud, in the capacity that it needed them. At the end of a recruits' course, the Adjutant asked one of the men who had passed out what had impressed him most. He replied that it had been the complete change of attitude in those very ordinary young men who, on first joining, had been irked by the sudden wave of discipline imposed on them. After the first week they had ceased to "moan" and by the time they were ready to pass out they had completely identified themselves within the Corps and took as much pride in it as a Sergeant-Major. The Commandant of the Depot was fond of repeating a conversation overheard between two passed-out recruits who were stood watching some new entries arrive. "Good God, Bill," said one, "Did we look like that once!"

There could be no higher tribute to the sheer hard work and seemingly endless patience of the Royal Marine instructors, many of them pensioners, who, besides instilling into the recruits the elements of drill and discipline, also passed on to the younger generations the tradition and high spirit of all members of the Royals.

The recruits passed on from the Depot to another camp, also in Devon, where they received instruction in platoon weapons, assault training, field-craft and battle-drill. The course ended with a 36 hours' tactical exercise in the country. Recruits who had been noted as potential officers at the Depot were followed up during this training, and those selected went before a Board after a small-arms course then passed on to the Royal Marines Military School. Marines selected for Corporals, and Corporals selected for Sergeants, had a six-weeks' course at the Deal Depot, where latent qualities of leadership were developed. Here again, potential officers were noted. Recruits who had technical qualifications were given a trade test at Fort Cumberland and, if successful, received further training. In this way the organisation made the best use of every man and sent out a continual flow of trained Marines to serve the land forces of the Corps, including its newest commitment, a force which was responsible for the ground defences of the Royal Naval Air Stations at home and abroad.

"Surprise was Complete"

The Royal Marine unit sent to the Middle East proved itself in the highly successful raid made on the wireless station at Kupho Nisi, a small island off the South-Eastern coast of Crete, on the night of 15th April 1942.

The raiding force, consisting of three officers and about 100 other ranks, led by Lieutenant-Colonel E.H.M. Unwin, embarked in H.M.S. Kelvin and arrived off the landing beaches at midnight. The weather was ideal; a calm sea and bright moonlight. The two arms of the bay stood out clearly as the force went ashore in the Kelvin's boats. Surprise was complete. The landing was unopposed. A party was left to cover the boats; another took up a position on some high ground near the beach to intercept reinforcements or to cut off the enemy fleeing from the station.

Four minutes after the assault platoon, under Lieutenant T.B. Heslop, had moved off, a machine-gun opened fire 300 yards ahead. Lieutenant Heslop, with the object of out-flanking the gun and obtaining a view of the objective, led the way to the summit of a rocky hill, then pressed on towards the station. There was no track and the ground was rough. During the approach the enemy opened fire again with a machine-gun and a few rifles. The fire was returned and silenced. Colonel Unwin, taking a Bren gunner, occupied a sandbagged emplacement on the left flank from which fire had been directed on the assault troops. It had a commanding position and, resolutely handled, might have jeopardised the raid, but the enemy had evacuated it.

The assault platoon reached the southern side of the station, opened fire on the windows, then raised a cheer and charged the building. Colonel Unwin closed the west side, followed by three Marines. Grenades were thrown through the windows and the Marines then forced an entrance – only to find the station deserted. The garrison had escaped by a door on the east side. Some paraffin lamps were still alight. Opened boxes of hand-grenades and ammunition were lying partially emptied. The wireless apparatus had been damaged by the Marines' grenades. Its destruction was completed. A large steel safe was removed and carried back to the boats, also two suitcases containing books and papers, and a mailbag of correspondence. Rifles were smashed and boxes of grenades and ammunition were placed in a heap and blown up when the Force was clear. This caused a tremendous explosion, which blew Colonel Unwin off his feet. The occupation had been made at the cost of only three casualties, caused by a bomb thrown from the roof by the last of the defenders, who had made off in the darkness. Surgeon-Lieutenant M. McRitchie, R.N.V.R., found the wounded men and attended to them.

Meanwhile, the reserve platoon, under Lieutenant H.H. Dyall, had moved forward to a pre-arranged position astride the track leading from the station. Voices calling in Italian were heard coming from what appeared to be a beacon fire a few hundred yards to the northward. The section sent to investigate discovered a small concrete structure, possibly an air-raid shelter. Having thrown grenades into the entrance, the section closed the place. In the entrance they found a large heap of inflammable material, which seemed to have been previously prepared and was now burning fiercely. More voices were heard, some swearing in German, from a party moving in the direction of the village. Owing to the rigid time limit imposed by the operation, it was not possible to go in pursuit, and the section rejoined the platoon to cover the withdrawal to the beach.

Carrying the stretchers over the rocks and down the precipitous hills proved a formidable task and the bearers finally had to pick up and carry the wounded men. But the enemy made no attempt to follow and the raiding force re-embarked in the boats without further casualties.

Within minutes the scene was deserted. They had all disappeared like shadows in tiger country. Back to their blue-water navy.

CHAPTER 15

THE ROYAL MARINES AT DIEPPE

The raid on Kupho Nisi, although not carried out by a Commando unit, closely resembled a Commando operation. That is to say, it was the type of perilous enterprise which the Royal Marines have been called upon to perform for the past 300 years; for although the term "Commando" was new at that time to the general public and, until the Second World War its role was an unusual one for the British Army, the Corps had long been familiar with its functions.

Pending the reorganisation of the Army after the evacuation of Dunkirk, the War Cabinet made plans to maintain a limited offensive in order to harass the enemy on the occupied coasts of Northern Europe from Norway to France. For this purpose small units, known as Independent Companies, were formed, and employed for minor raids and reconnaissance's. All the officers and men were volunteers. Such raids were the very tasks for which the Royals had always been held in readiness, either embarked in ships or at land bases. Their organisation in every aspect of fieldcraft and training had been dedicated to that end.

At the time when the need of these small units arose, however, the only Royal Marine assault force not afloat in H.M. ships was the Brigade itself, which was being held as an amphibious force to meet an emergency which was expected to arise at any moment. Overall, it would have been impossible for the Corps to have undertaken the formation of small raiding parties without breaking up the Brigade, or at least leaving gaping holes within its ranks, and since the detachments afloat could not be withdrawn without depriving the ships of trained gunners, the independent units had to be drawn from the Army.

These Army Commandos, as they came to be called, increased so fast that it became necessary to administer them under a separate headquarters known as the Special Service Brigade. This was an entirely military organisation, under the control first of Lieutenant-General (later General Sir) Alan Bourne, then Adjutant-General, Royal Marines, and later of Admiral of the Fleet Sir Roger (later Lord) Keyes. Eventually the Special Service Brigade developed into the organisation to become so well known as the Combined Operations Command, under Vice-Admiral Lord Louis Mountbatten, the forces employed being drawn from the Royal Navy, the Royal Marines, the Army and the Royal Air Force.

Early in 1942, however, before Combined Operations Command came into being, plans were made for the inception of a small raiding force which could be held at the disposal of a Naval Commander-in-Chief. It was to be carried in H.M. ships or kept in readiness at a Naval base. It was to be known as the Royal Marine Commando, but was not to form part of the Special Service Brigade. The theme of the day was well-trained, hard-hitting independence which could prove flexible with logistical support and not fall at the first fence under the first signs of hardship or deprivation. To my understanding, what control and command were looking for here were men who, when confronted with a God-awful situation could, and would, always fall back on the old tried and trusted adage "When the chips are down and stormclouds have gathered, drop the plan and improvise". Create, and let creativity move you forward at all cost. But above all else, don't just give up. The advantage of using only trained Marines for such a purpose was that they were familiar with boat-work and were used to the sea. Members of the Corps who were above average shots and not liable to seasickness were accordingly invited to volunteer for "Special Service of a Hazardous Nature". That very official phrase, which today has been cut down to "Special Service Duties", and of which I am so - to use that word again - "familiar" with, discloses so little but suggests so very much, proved sufficient inducement, and a force some 450 strong was selected from the numerous applicants, about equally divided between men enlisted for Continuous Service and for Hostilities Only.

This new-born Royal Marine Commando Unit was originally allotted a special mission against the Japanese lines of communication in the Pacific under the American Naval Commander-in-Chief in Far Eastern waters. By the time it was fully trained, however, developments in the Far East led the Admiralty to believe that the plan was no longer feasible, and it was then decided to use the Commando in the Dieppe Raid.

I would wish to tell you a tale or two on those would-be Far Eastern missions for the new Commando, but try as hard as I did to dig out the finer details of what was planned with the Americans, I was surprisingly "blanked" on the subject and advised not to pursue the matter. Needless to say I complied with their wishes, but as a former "member" as they say, and with the greatest respect for the Far Eastern theatre, I can't but wonder at the stories we've missed and thc books that have never been written had there been a real fight for Singapore. I know David Lean would have won more Oscars if he had found a chance to put that on film. But, it was not to be, so the fortunes of war took us to Dieppe with, who else! None but the brave.

"With a Courage Terrible to See"

It was first given a purely Naval role: to seize the coastal craft and barges in Dieppe Harbour and sail them back to England. Untoward events made a change of plan necessary in the initial stages of the raid, and the Commando was then detailed to land on White Beach, the centre of the assault, and reinforce the Royal Hamilton Light Infantry. I will not dwell on this operation as it has been described in fine detail in another book "Combined Operations", including the story of how Sergeant T. J. Badlan, R.M., of the Beach Provost Party, won the D.S.M. by taking charge of a landing craft when all the crew had been killed or wounded. Although, to quote the official report, "'With a courage terrible to see', the Marines went in to land determined, if fortune so wished, to repeat at Dieppe what their fathers had accomplished at Zeebrugge," the Commando did not achieve its object and suffered heavy casualties, including the loss of its Commanding Officer, Lieutenant-Colonel J.P. Phillips, who, by putting on a pair of white gloves so that his hands could be more easily seen while directing the movements of the landing craft, saved 200 of his men from entering a field of fire which must have proved mortal to many of them.

After the Dieppe Raid, the Royal Marine Commando was reinforced and a second was formed, the nucleus being provided by three companies from a battalion of the Royal Marine Division. These units, each some 500 strong, came under the operational control of the Chief of Combined Operations, who, as a Naval Officer, had a high appreciation of the value of Marines for amphibious operations.

Each Royal Marine Commando consisted of a headquarters and six troops. The troop commanders were relieved of as much administrative work as possible. "They have nothing to do bar fighting", as one of the Adjutants observed. Each Commando unit was entirely self-supporting, with its own cooks, butchers, shoemakers, blacksmiths, motor mechanics, carpenters, armourers and equipment repairers sometimes known as sailmakers. Normally these specialists were at Base Headquarters, but all were trained fighting men who at a moment's notice could down tools and compensate for any role in a front line theatre of war. There was also a proportion of officers and other ranks who were trained parachutists; and at this point in our story I think it is fitting to mention that a Royal Marine officer, Lieutenant-Colonel R.G. Parks-Smith, who was mortally wounded at Dieppe while in command of the Beach Provost Party, was one of the pioneers of parachute descents in the Second World War.

The average age of the men in the Royal Marine Commandos was between 23 and 24; and over 90 per cent were those enlisted for Hostilities Only. Looking back on those early, vital but meagre days, to understand the knife-edge life they had to lead, one has only to make a comparison between the high-tech efficiency we watched on T.V. during Gulf War Two, in relation to the time of Dieppe when they had no service rations or service quarters, but were paid a special allowance to cover billeting and food, so that they could be entirely mobile. In the field they would carry composite rations. They wore khaki, with the "Royal Marine Commando" flash on their shoulders, and green berets with the Globe and Laurel badge. Their training, which was similar to that of the Army Commandos, made them tough, but this particular quality was discouraged when they were not on duty. Their officers made it clear that their fighting abilities were expected to be apparent only in the field. As in the other units of the Royals which served ashore, the use of sea terms was encouraged and was readily accepted by the men. Once again, this has a psychological effect. As members of a Corps which had its roots in the Royal Navy, they felt it a point of honour not to be sea-sick when they were on board, or to put it another way, afloat, in however small a craft.

There were other officers and men, both from Royal Marine Commandos and acting independently, who showed conspicuous gallantry in the special services for which they had volunteered. In one document I read, the statement was simple and to the point; dated 1943, "Their activities are still secret and likely to remain so until the war is over. At present it is but possible to give an indication of the hazardous enterprises on which they embark, working alone or in small companies, unsupported by any covering force, but relying upon their own initiative, their own sharp wits, and their high courage, to bring them through".

Just one of these actions which I know we all remember from film was the story of "The Cockleshell Heroes" with of course, Bill Sparks, D.S.M.

"The 'Winkle Barges' at Dieppe"

Closely associated with the Commandos, and operating under Combined Operations Command, were the landing craft whose armament was entirely manned by Royal Marines. Their task was to accompany the first flight of a landing force, to close the beach, and to destroy at close range the enemy's forward positions and lighter guns. Behind them would come the assault craft carrying the infantry, followed by the tanks, and then by the landing craft flak, whose job it was to put up a close-range barrage over the whole party. Other craft were available to tackle the defences in rear, or any targets which proved too formidable for the support craft. Then would come the barges with the main body of the infantry, and other barges with heavier armament, the cruisers and battleships covering the invading force offshore. Each support craft was commanded by a Sub-Lieutenant, R.N.V.R., with an engine man and four ratings. The armament, which may have been a 2-pounder and an Oerlikon, was manned by a Corporal and six or seven sea-service Marines, enlisted for Hostilities Only. They had their initial training at the Portsmouth Division and proceeded for further instruction in handling their craft to what was once a seaside holiday camp, where they came under the control of Combined Operations Command. The landing craft flak carried an R.N.V.R. officer as navigator, also engineer and deck ratings, and two Royal Marine officers and some 45 Marines, who would man the armament. These flakships, or "winkle barges", as they were affectionately called, went into action for the first time at Dieppe, where they proved their worth by covering the assault landing craft and "buzzing up and down the beaches", as one of their officers described it, giving close-range fire to the troops as they landed or after they were ashore by shelling pill-boxes and machine gun nests. They destroyed a number of German aircraft, one alone accounting for no less than eight. This is a description by the officer commanding the Royal Marine detachment which shot down the flakships' first victim:

"At 0700 we received orders to proceed at once to White Beach, and give close support to our troops who were being held up by two white houses. As we moved up we received our first attack – two ME.109's in line ahead, flying directly in from the stern. The attack being at low level, only the two bridge Oerlikons could bear. When the leading Messerschmitt was about 800 yards astern, the starboard gun opened fire, the other gun taking on the second aircraft. The effect was electrical. The very first burst scored direct hits on the leading Messerschmitt and it wavered, dropped and rose again rapidly, then it just evaporated. One minute an aircraft, then 'wallop', the next minute a high black cloud and just hundreds of small bits tumbling everywhere. The very aviation fuel that kept him in the air was now instrumental in his destruction. The second aircraft was also hit, climbed steeply, and passing slightly to our starboard was engaged by the for'ard Oerlikons and starboard pom-poms. It was severely hit, but disappeared into the thick smoke before we saw it crash. The whole action only lasted a few seconds, and then pandemonium broke out on board. The guns' crews were cheering madly and dancing around shaking hands, slapping each other on the back – in fact, the action was temporarily forgotten. It was the first time we had ever fired against enemy aircraft and we had one certainty and one probable."

Another flakship, having shot down two JU.88's, came out of a smokescreen 400 yards from one of the beaches to be attacked by heavy machine-gun fire from the shore and by nine German fighters flying low in close formation.

"It was a most unnerving moment," wrote the Royal Marine officer in command, "To see the orange and yellow tracer tearing all around and passing you, or even coming straight at you but for some reason never hitting home; it appeared to come so slowly but then pass you like so many lightning flashes. But the guns' crews stood up to it magnificently and let go right into the middle of them, and although I did not see one crash they could not face our barrage and broke off in all directions."

During the raid, and after the troops had been withdrawn, the flakships picked up a number of survivors from the water, many of them wounded. Some were from the Royal Marine Commando. "One Marine still had his Bren gun," wrote the officer first quoted, "And he was no sooner on board than he demanded oil and flannel, and lighting a cigarette began to clean and dry his Bren." The same Marine was later heard to complain bitterly when he discovered he had lost a magazine for his Bren and a barrel pull-through which he had used constantly for over three years. When I read this it struck me as strange how you become attached to the simplest of things in times of conflict, but I understood his loss as the same thing happened to me in Ireland. (Small digression here).

I had a marvellously sturdy Mils Prismatic Compass which I had "acquired" during the close-down and withdrawal with 40 Commando from Singapore. It was my constant companion in the field for over seven years with just two breaks for calibration and fluid change. Then one hot August night during a long arduous trek from the badlands of South Armagh to the back-door approaches of Belfast by the great Black Mountain, I stopped by a desolate tumbledown old cowshed to take stock of my position. A leather holster on my belt which housed a 9 mm Browning Automatic also had a small canvas pouch next to it in which the compass fitted perfectly. On reaching round to check the weapon (a frequent force of habit when on the move), I realised the compass was gone. The small press-clip had come undone and the compass was just gone forever. On the terrain I had crossed there was no hope of going back without getting caught in bandit country at first light. The pouch had come undone and I had lost it through the stupidity of not paying closer attention to fine detail. Back in basic training or at J.W.S., Jungle Warfare School in Malaya, I can remember one of two P.W.'s, Platoon Weapons Instructors, who would have all but had me shot for this, so in a way I can understand the Bren-gunner's aggravation at his loss.

Later my only consolation was remembering that scene in "Laurence of Arabia" when he discovered he had lost his compass while crossing the desert. I just got that feeling, well, if it was OK. For T.E. to lose his compass on his way to Cairo, then surely it would be O.K. for me on my way to Ballymurphy! Although for the life of me I just couldn't imagine the likes of David Lean or his successors rushing to make an epic film about it.

Now to get back to Dieppe, the raid was but a preparation for invasion, and both the flakships and the more lightly-armed support craft were employed in the first landings in Sicily, where they gave a good account of themselves, the support craft going right in to the beaches with the first flights of the assault troops. The numbers of these craft were increasing fast, and as the invasion of Europe progressed more would be heard of their activities. The initial success for landing on enemy territory depended upon the efficiency and initiative of those corporals and their gun-crews of Marines, and no greater compliment could have been paid to the Corps than to entrust them with the great responsibility they were to bear.

CHAPTER 16

SPEARHEAD OF INVASION: THE MARINES IN SICILY

For many months before the invasion of Europe began, the Chief of Combined Operations had, in the two Royal Marine Commandos, a weapon waiting to be used when and where its striking power and trained durability would be most effective. If you can look back with your imagination and try to view him like an archer with a bow whose string had never been fully extended; but his arrows had been skilfully fashioned and were ready in the quiver. The target was known only to himself and to others in the Higher Command. Those banded brothers in the Commando Units would not know it until after they had embarked in troopships at the end of June 1943, and found themselves steaming in convoy to the southward.

Then, and not till then, did they learn the meaning of the operation they had been rehearsing in Scotland for so many weeks, and the end of speculation came with the announcement that their destination was a sandy bay in southern Sicily. Their task was to assault and silence the coastal defences of the main beaches before the 1st Canadian Division was put ashore. No mission could have been more fitting for Royal Marines. Now, while I am not one for bandying names about, I will simply tell you it had been assigned to them by one very important person who was not unmindful of the high achievements of the Corps.

Once the secret had been revealed, maps and air photographs of coastline were minutely studied. Every detail of each troop's task was reviewed in the light of the latest intelligence reports. Anyone who read "The Fourth Province", (some people refer to this story as speculative fiction on future events in Ireland, on that we shall see), will know the voracious importance I repeatedly place on good intelligence. The Int. Factor would be vital to success then, just as it is now in the 21st century. The Company and Unit Commanders knew that good, fresh intelligence was worth three pounds per pound when first caught, but just like fish, three days old and you can give it to the cat. There was the narrow beach – Commando Cove, they called it – in the most southerly tip of the island, backed by low cliffs midway between two headlands, Punta Ciriga on the west and Punta Castellazzo on the east. One Commando, led by Lieutenant-Colonel B.J.D. Lumsden, would land first and scale the cliffs. The leading troop would mop up enemy resistance on the top and form a bridgehead behind which the main party would organise before pushing forward. The remaining troops would move off to the right, clearing machine-gun nests and pill-boxes along the cliff and assaulting the enemy's posts on and behind Punta Castellazzo which might prove an obstacle to the Canadian's landings. Finally, they would destroy what appeared to be a formidable strong-point which covered the western end of the main landing beach, and would make contact with the Canadians.

The Second Commando, led by Lieutenant-Colonel J.C. Manners, was to land in the same cove ten minutes after the first, pass through the bridgehead and move to the left, attacking the coastal defences as far as the village of Solarino, a mile and a half to the west. Once the coast had been cleared, both Commandos would consolidate their positions and hold the left flank of the Canadian beach perimeter. Alternative plans were made in case a landing in Commando Cove proved impracticable, or the ship carrying one of the Commandos should be sunk, when either would be prepared to take on all the tasks allotted to both.

These plans were explained to every Marine who would be taking part in the operation. Each man knew exactly what he had to do. All expected resolute opposition and were prepared to overcome it. Nothing was left to chance.

"On the Night of 9th July"

On the night of 9th July, the great invasion fleet drew under the lee of the Sicilian coast. There had been a stiff gale earlier in the day and the sea had been running so high that at one time it seemed doubtful whether the landings would be possible. After sunset the wind dropped, leaving a heavy swell.

Just before we move into the action, I must mention here the details I managed to glean from numerous military sources on these actions were sometimes, although accurate, very rigidly reported and scrappy, which can make for a rather boring series of fact-finding statements. Unfortunately, this does nothing to bring to the surface the real feelings of being up-front with the men in the boats. So, familiar as I am with Mediterranean beach landings, let me set the scene for you as I imagine it would have been on this night in July.

The night was clear; the moon in its first quarter. At midnight, when the troopships were lying some eight miles off the coast, the Commandos embarked in their landing craft. The flotilla formed in two columns and headed for the shore. They could feel the heavy swell under their boots and could taste and smell the proximity of the clean, salty water.

As the moon went down, a new light sprang up on the horizon; flares which the R.A.F. were dropping over Pachino, the town north east of the beaches. Their brightness silhouetted the coastline ahead. Then followed the flash and thud of bombs. The R.A.F. had undertaken to light up Pachino for the assault troops. They were keeping their promise.

The flat, black assault craft drove swiftly forward, leaving a luminous wake on the dark water. Engine trouble compelled one of them to turn back, with half a troop of Colonel Lumsden's Commando enraged with disappointment. The following seas threw up the sterns of the remaining craft and caused their square ramps for'ard to dip perilously low. Spray drenched the tightly-packed Marines in the well-decks. The craft began to ship a good deal of water. One took two heavy seas over her ramp. The steel doors for'ard were burst open and the water went pouring aft. The well-deck began to fill alarmingly. Some of the Marines clambered onto the gunwales and the flat top of the engine-room so that their comrades might have room to work the hand-pump and bale with their steel helmets. Nevertheless, the water continued to rise slowly, until it was knee-deep.

By the time the flotilla had closed the coast, the high seas had thrown it somewhat off its course. Steaming barely 200 yards from the shore, the assault craft skirted the cliffs for nearly a mile, searching in the darkness for Commando Cove. This was the tensest moment. Would they find the beach before they were seen! Or were the enemy watching their advance, ready to open fire! Support craft astern of the flotilla were prepared to give cover, but held their fire in the hope that surprise would be achieved.

The flotilla steamed on, expecting at any moment to be shelled by the coastal batteries. Suddenly there was a report and a gunflash from the cliff. Instinctively every Marine ducked his head. But it was an anti-aircraft gun engaging the R.A.F. bombers overhead. Then the flotilla turned into the dark embraces of Commando Cove and deployed into line abreast for that last breathless dash to the beach, apparently still undetected.

While the flotilla was racing inshore a solitary machine-gun opened with a short burst, then was silent. The assault craft sped on. As they grounded, the same gun fired a second hesitant burst. Silent and unhurried, the Marines of Colonel Lumsden's Commando jumped from the ramps into the waist deep sea and began to wade ashore, holding up their weapons, each with his allotted load. Some carried ammunition, some equipment, others cans of drinking water.

As the dark mass of men were struggling through the gleaming surf, the machine-gun on the cliff was strangely silent. Not a shot came from the position known to be on Punta Castellazzo, the eastern headland. Beyond the sandy beach, instead of the expected cliff, was a rough shelf of limestone, over which the Marines were able to scramble with both hands full.

"The whole process seemed fantastically deliberate and leisurely," wrote a Royal Marine officer who accompanied the Commando. "It was hard to believe that this land beneath us, the first we had trodden since leaving Britain, was enemy soil, the first bastion of the fortress of Europe."

Then the machine-gun on the left opened up again, and this time there were a few casualties. Without hesitation the Lewis guns of the assault craft engaged it.

"The Bridgehead is Secured"

The touch-down had been made within 100 yards of the point originally planned. Landmarks that had been well memorised from the maps and photographs were picked up and each troop moved off to its appointed task. Soon, from an isolated building on the immediate left, came the sound of a bursting hand-grenade, rifle-shots, and cries in Italian. A swirling red stream of tracer from the Marines' Bren guns shot towards the post, some of the bullets ricocheting high into the air off the rocks and shale of the emplacement. There was little more trouble from this post.

As soon as the bridgehead had been established, Colonel Manners' Commando was signalled ashore, landed, passed through, and turned left-handed on its mission. Bursts of fire on the right and brilliant streams of tracer showed that Colonel Lumsden's men were engaged. But the resistance was irresolute. The Italians would fire a few bursts from their rough stone emplacements, thatched to resemble farm buildings, but surrendered when they saw the Marines at close quarters. Step by step the defences were silenced, until the attackers reached Punta Castellazzo. There was a sharp engagement with a wired-in strongpoInt. Then the defenders raised their hands. As the Marines closed in, one of the Italians threw a hand-grenade. "The sequel," observed the officer already quoted, "was not one to encourage any repetition of these tactics."

A group of buildings occupied as barracks, was cleared and prisoners taken. The remaining section of the troop which had been compelled to turn back to the ship accomplished successfully the task which had been allotted to the whole troop; exploiting eastwards, they finally joined hands with the Canadians, who by that time had landed and were pushing inland from the main beaches. Two other troops attacked a strong point known as Casa Della Marza, which commanded the western exit from the Canadians' beach – the road from Pachino to Spaccoforno. The main building was cleared without difficulty, but two machine-guns opened from corners of the enclosure. Both surrendered when the Marines began to advance for the assault. A sniper was dislodged from a low tower surmounting the Casa.

Meanwhile those in the support craft and the destroyer lying close inshore watched the green success lights going up from point to point as Colonel Manners' Commando stormed its way westward, until the coastal strip as far as Solarino had been cleared of the enemy and the left flank of the main bridgehead secured. As the sun rose, the whole area between Solarino and Punta Castellazzo was firmly in the hands of the two Commandos. Landing craft bearing more Canadian forces were streaming towards the beaches, with no opposition. Overhead a solitary flight of Spitfires demonstrated the R.A.F.'s mastery of the air.

Patrols of Marines and Canadians pushed inland along each side of the road to Spaccoforno, occupying the high ground to protect the beach perimeter. So completely had the defences been overwhelmed that by 7 a.m. an officer was able to ride an Italian bicycle down the Pachino Road in perfect safety. The civilian population, peasants of the poorest class, seemed to have unhesitating confidence that they could go about their work unharmed. They smiled, touched their caps and offered thin brown cigarillos to the passing troops. They raised no objections when their brightly-painted carts, drawn by gaunt-looking mules and donkeys, were commandeered and loaded with mortars and ammunition.

A cruiser and a monitor, lying close inshore, were administering a tremendous pounding to the defences far inland. In the sunlit bay a great mass of shipping lay at anchor, while the landing craft plied busily to and fro, and amphibious lorries rolled straight out of the water and up the beach, without interference from the enemy. All this had been accomplished within five hours.

The Royal Marine Commandos had been allotted an historic mission. They had been chosen as the very spearhead of the invasion so long awaited by the world. They had taken the pain, fear and isolation of the first probes into an alien enemy land where awaited – they knew not what mayhem and catastrophe for would-be trail-breakers. They had been the first sea-borne troops ashore. The confidence and belief in their ability of that great man in London had indeed been well placed. They had done all that had been asked of them. They had destroyed the coastal defences which might have held up the Canadian Division. They had cleared the way for those who were to follow them. They had held the left flank of the main landings against possible counter-attacks; there had been no allied troops between them and the Americans at Gela, 50 miles to the west. Even had the opposition been fiercer, it is not likely they would have failed.

I never knew an appeal to them for honour, courage or loyalty that they did not more than realise my expectations. If ever the real hour of danger should come to England, they will be found the country's sheet anchor.

Lord St Vincent

After the occupation of Sicily the two Commandos took part in the first landings in Italy itself. During the night of 7th September Lieutenant-Colonel Manners' Commando secured a bridgehead at Vibo Valentia, in face of opposition from enemy mortars and 88 mm. guns, and captured the town. Later they made another landing in advance of the Eighth Army at Termoli on the Adriatic Coast. On 9th September, Lieutenant-Colonel Lumsden's Commando landed with the Fifth Army at Marina, to the west of Salerno. In company with an Army Commando, they took the town of Vietri and seized La Mollina defile, through which ran the shortest road to Naples. This defile they held against determined German counter-attacks for five days until they were relieved by the main forces advancing towards Naples.

During these operations, the two Royal Marine Commandos had performed tasks which were wholly in keeping with the traditions and functions of the Corps; their forebears had made just such landings at Gibraltar and Belle Isle, at Gallipoli, and on a hundred other beaches all over the globe. In the future their Silver Bugles would have yet another day – the 10th of July – on which to sound a fanfare of honour.

What I have told you about the Royal Marines is, I consider, a reasonably fair selection of events which I managed to dig from the finely detailed memory banks of the Ministry of Information, H.M.S.O., and a wide selection of well-informed institutions. So, when is enough, enough! Well, from the great volume of information I was offered, there was always a danger that by including most events I might end up with a book of house-brick proportions. A complete monstrosity to rival War and Peace, or Joyce's Ulysses. God forbid, for this was not my poInt. I think most people these days want to read a history of their chosen interest with a sense of ease that does not burden or test their patience, but rather leaves them with a satisfied understanding of what – why – where - when and how. I want to keep brevity and, as near as possible, 'sharp detail' as my watchword on this book, and by doing this I might ensure it is not discharged in misunderstood frustration from too many train windows. I would also hope it could be found somewhere amongst the kit of every soldier's locker, especially Marines and certainly the Royal Navy.

By reaching away from the Second World War I would like to bring you up to date on some recent events and allow you to judge the huge volume of progress the Corps has made, not only on the home-grown problems of Ireland, but also the sheer flexibility of command and control displayed during the Falklands War.

So, without further ado, let us move forward in the time machine, "saddle-up", as it were and get back into the field around about August 1969.

"Gentlemen," he announced, "The troops are going into Belfast."

General Sir Ian Freeland
at a London press briefing
Friday 15th August 1969

CHAPTER 17

A STATE FORGED IN CONFLICT

"THE TROUBLES"

This is the term given to years of conflict in Ireland which has seen thousands dead, widespread destruction and horror upon horror too often unreported. It can often be a contentious issue of debate of when and how Northern Ireland's "troubles" began, who and what is to blame. You can go back 35 years, or even 300 and beyond, for in reality Ireland has been engaged in conflict with England for many centuries.

It could well be described as Northern Ireland's forgotten conflict. The closing years of the 1960's will be remembered in history as a period "when the road to democracy opened up". The march for civil rights and equality had come to the fore in America among the black population, especially in the Southern States. Disillusionment with the Vietnam War began to over-spill as television images brought the war into American households. In Paris and London, C.N.D. and anti-war rallies turned into clashes with the police as another generation took its views and policy for a better world onto the streets. Rightly or wrongly, the sixties was making its mark as a period of change, and as that decade came to a close it was a change that would leave society facing a rapidly-changing world.

In the tiny state of Northern Ireland, the drive for civil rights came very much to the fore in 1968, causing an insecure Stormont Government to over-react to such an extent that an overspill of resentment and frustration resulted in rioting on a scale never before experienced. It was not long before Westminster knew that troops were inevitable. The Ulster Commander, General Freeland, had a senior officer in civilian clothes on the streets of Derry; he reported that the R.U.C. could not possibly contain the Bogside area of Derry for more than thirty-six hours, such was the force of unrest. When they retreated, he said, there would be considerable blood-letting between the communities.

Quietly, Freeland, based on this report, sent 300 troops to the Sea Eagle naval base on the Waterside close to Craigavon Bridge, ten minutes drive from the centre of Derry. It was these troops, men from the First Battalion, the Prince of Wales' own, who were deployed into Waterloo Place at tea-time on Thursday 14th August 1969. History was in the making as these young British soldiers marked the beginning of what has now reached 35 years of British Army deployment on the streets of Northern Ireland.

Thinking back now, I remember watching these events coming to the fore on every news briefing on television. I was then in 3 Troop, 'X'-Ray Company, 45 Commando Royal Marines at Stonehouse Barracks in Plymouth. We were in the throes of packing everything up at Stonehouse and moving to our new location in Arbroath on the east coast of Scotland to take on the Arctic role in defence of the northern flanks of N.A.T.O. But events in Ireland were to interrupt our Arctic role on many occasions over the years, and over six tours of duty I was always deployed with 45 Commando. Only two of these four month tours were in uniform on the streets, the rest were in the area of intelligence work. Much of the intelligence work that has been carried out in the province undoubtedly comes under the Official Secrets Act and cannot be told for some considerable time. There are, however, certain things which can be revealed, so I must be selective on the stories I tell on the Irish section of this book.

Over the years a fairly large intelligence section was set up and grew considerably as the troubles began to escalate. It was set up at Army headquarters at Lisburn, Co. Antrim. When the Army first arrived they knew little about the local I.R.A. or Protestant extremists, and had very few contacts. They relied on the Royal Ulster Constabulary for good local knowledge. Some of the information they were given from R.U.C. sources on intelligence was so far out of date that many of the people who were picked up when the initial internment operation was mounted had not been active in terrorism for years.

In those early years of trial and error the key to success in internal security operations would always be good intelligence. On my first operations with 45 Commando in Belfast good intelligence came from two levels – agencies outside the unit and from information gathered by the men on the ground. In my particular area of operation, the unit would maintain close contact with R.U.C. Special Branch working within their Police Division and, of course, with the Brigade Intelligence Section which would pass source and interrogation reports and intelligence summaries to us. Within this the unit Intelligence Sections were set up. For me, these initial stages of intelligence gathering created what I can only describe as a monumental nightmare of paperwork. We had to organise a local census in our area of operation so that street registers could be built up and records kept of all known I.R.A. men and women, their families, associates, their habits and way of life. Those who were wanted were taken to headquarters where they were screened by the intelligence officers. This involved taking photographs, establishing their identity and recent movements. Those thought to be involved in terrorist activities were sent to a Special Branch interviewing centre, where they were questioned in depth.

So, by means of source, interrogation and intelligence reports, checks, screening and the inquisitiveness of the Marines patrolling the streets, we could build up files on the I.R.A. units, photographs of wanted men and their associates, details of safe houses, meeting places, and the likely locations of weapons, ammunition and bomb-making materials. The bottom line was, based on good intelligence, operations were mounted to arrest those on the wanted lists and if possible recover arms, ammunition and explosives.

My group, 'X' Company 45 Commando, were frequently involved in rapidly planned and executed cordon and search operations which often brought about violent aggro situations as the local women and children would stage riots to provide diversions while activists uncovered and moved weapons of all sorts which would have been hidden. Specially trained search teams are always on call for house searches where dogs and mechanical sniffer devices could be used.

In addition to house searches, 'X' Company had to check on derelict buildings and waste ground for hidden weapons, ammunition and explosives. During all of these operations the teams needed to be fully alert to the dangers of command-detonated mines or booby-traps. In an effort to inhibit the actions of known or suspect individuals dealing with arms and explosives, vehicles were searched at random. Normally, snap V.C.P.'s or Vehicle Check Points would be mounted and remain in position for about 10 minutes. However, when car bombs became a real threat, we carried out hasty searches of large numbers of vehicles for canister or cylinder bombs.

During their tour of duty from April 1973 to October 1973, 40 Commando Royal Marines searched over 115,000 vehicles.

Almost every commercial establishment had by now installed their own security systems. Bar and shop owners placed barrels or concrete blocks outside their premises and had security staff on constant watch for car bombs, or rolling ambushes with hit and run gunmen. It was now the policy of every Unit Commander, from Brigadier C.S. Wallis-King at 3 Infantry Brigade in Lurgan, down to the level of Section Commander in a Rifle Troop that the task at hand was the neutralisation of any extremist organisations involved in terrorist activity. This would include Loyalist and Republican terror elements as well as the Provisional I.R.A. Depriving these illegal organisations of their weapons would go a long way towards obtaining and maintaining an acceptable level of violence on the streets. But from those very early days we all understood the problems of recovering illegal ordnance was a skill to be worked at and learned. Trial and error was the watchword, but as successful as we were there was far too much error. Even today, over thirty years on, the Provisional I.R.A., although still - at time of writing - on cease-fire, have unhindered access to some of the finest small-arms anywhere. In 45 Commando, no stone was left unturned in digging out information on the art of weapon recovery.

CHAPTER 18

SEARCHING AND FINDING

"Attempt the end, and never stand in doubt;
nothing's so hard, but search will find it out."

- (Herrick: Seek and Find)

The "troubles" in Northern Ireland have always been a bitter form of tribalism, and no amount of window-dressing or make-believe will help to make it something noble or decent. In order to maintain its required level of terrorist operations, the I.R.A. and other illegal organisations from the early 1970's and onward, required the permanent presence of arms, ammunition and explosives readily available for immediate use, either for bombing or shooting. It was a fact that a great deal of terrorist equipment was hidden in hundreds of hides throughout Northern Ireland at that time. Some were just a few days old; others had been there for many years. They were in fields, bushes, embankments, derelict houses, disused quarries and many other places. The hides varied in size from a saucepan without a handle to a small hut. Experience taught us that all hides or sets of hides discovered, although in different parts of the province and hidden in many different locations, had sets of factors which were common to all. The task for 45 Commando and all the Security Forces in searching for these hides was daunting and would remain so if the correct research and deductions from that research were not made. What the Security Forces had to become capable of doing was mounting searches, in the absence of information, acting solely upon deductions drawn from the factors. That this could be done had been underlined by the record of two tours of routine patrols/searches in which ‘3’ – Royal Regiment of Fusiliers, had been involved.

During two tours of duty a Platoon Commander and his men had found 24 terrorist weapons, 4,420 rounds of ammunition, 37 detonators, 1,600 ft. of Cordtex, telescopic weapon sights, radios and explosives, in a total of 63 hides. The problem for the terrorist was to hide arms, ammunition, explosives and warlike stores in readily available hides for immediate use in the continuance of terrorism. But due to the activities of the Security Forces, the inquisitiveness of children, and the risk of chance discovery by passers-by, it became apparent to them that to hide equipment in their immediate surroundings was too dangerous and therefore they had to look elsewhere. They had to find an area or place which afforded both concealment and accessibility. It was the careful balance of these two factors which was their primary problem, and to achieve it they had to sacrifice a little of each. “Where and how could they conceal a weapon?” was the double question continually asked.

As far as concealment was concerned there was no shortage of places. They had woods, thick hedgerows, and hundreds of derelict buildings all miles from anywhere, which the Security Forces never went near. But if they used these places, where their equipment was perfectly safe, instead of hiding areas much closer in and more likely to be found, they would lose accessibility. (A weapon hidden in a South Armagh border region is no use to a sniper in Belfast). It was this accessibility factor which worked in favour of the Security Forces, for it turned out the terrorist who operated and lived in the town or estate must be able to reach his rifle or explosives quickly and easily, whether on foot or in a car. As far as concealment went, he had to be able to move to and from his hide, open it and take out his equipment and later return it without any risk of being seen. On balance he had to go for accessibility.

Terrorist weapons were issued to individuals who would be then responsible for their individual concealment in a place readily accessible by day or night. Sites would be near target areas so that there was the minimum carriage between the hide and the scene-of-use and resultant need to avoid Security Force patrols, vehicle checkpoints, or surveillance. The Royal Marines found in many housing estates bordering on the countryside terrorists would use two and often three main sets of hides:-

Storage Hides: Over 1,000 metres out and generally sited on deserted roads/lanes. The stores hidden in these hides would normally be there for long periods of time.

Transit Hides: One or two fields out. The weapons/stores would remain in these hides for two days to a week. They would be brought in from the storage hides when a job was in the planning stages.

Inner Hides: Sited within the estate in varying locations under paving stones in the darker, more secluded paths and alleyways, or in the bordering hedgerow, the main requirement being that the weapon could be back in its hide within at least a minute of the shooting. The weapons could then lie in the inner hides for a number of days until the "heat" had cooled off, when they could then be transferred back to the safer outer hides.

Terror groups found the best time to move their stores from one set of hides to another was between 02.00 and 06.00 hours in a time of minimum Security Force activity. Once their weapons were in the transit hides they could then easily be reached on foot and brought into the housing estate at any time of night in relative safety.

The Security Forces would be much more of a threat to terrorist stores if they initiated an immediate follow-up search after every shooting incident, even at night. All the S.F. would need to do would be to lift every paving stone that could be raised by hand in secluded alley-ways. This was proven by one case in point when the I.R.A. lost a Thompson sub-machine gun and an Armalite rifle only 180 metres from where they had been fired, all because the S.F. initiated an immediate follow-up search. The same I.R.A. unit some weeks later again lost two important weapons within 200 metres of the place where they had killed a policeman the night before. These were hidden underneath one of the few paving stones that could have been raised without any effort by hand.

Research and planning was found by far the most important part of any search operation for unless it was done thoroughly and logically, many man-hours would have been wasted searching areas in which no terrorist in his right mind would have hidden anything. The Marines learned many valuable lessons during these early years of trial and error "policing" in Northern Ireland, and learned many unique methods from the training and hard work put in by the Royal Regiment of Fusiliers.

CHAPTER 19

40 COMMANDO BELFAST 1972/1973

WE ALWAYS KEPT "ONE FOOT ON THE GROUND"

Our latest records show the Royal Marines have served around 41 tours of duty in Northern Ireland. I will now, over the following chapters, try to condense thirty years of field experience down to a fair and typical selection of incidentals with advice received from colleagues in 40 Commando who helped with stories for David Barzilay, as stated in acknowledgements, (The British Army in Ulster Vol. 2 - Century Books). As for the stories on 45 Commando who were responsible for Police Division H, I speak here with authority having been a fully operational "loose cannon", as I was described by certain R.U.C. officers. I think this had something to do with the fact that I was born, raised and schooled in the magnificent Drumcondra, the "Beverley Hills" of North Dublin, and after almost 45 years in the U.K. I can still fall back on the accent at any time. This for me was a very useful weapon to use if I found myself in any sticky situations in Republican areas. But then again, it caused much alarm and despondency within the bars and canteens of R.U.C. stations which I frequented on an almost daily basis.

The fact I was wearing civilian clothes and was on secondment from 45 Commando cut no ice at all with some of the old hard-line peelers (police) from R.U.C. Special Branch, and this they made plain to a point where my Company Commander (whom I shall not name) warned me to stop using my "Terry Wogan" voice in establishments of Loyalist influence. Even the deep brown skin of our Indian tea-wallah at Bessbrook Mill would turn a lighter shade of pale when I spoke to him with a southern accent, but these were tense and desperate days in "bandit country" and I must admit I did have a wild sense of humour, which is probably why they dubbed me a "loose cannon".

Before I cover my above tour with 45 Commando, let me give you – just as a comparison between urban and rural ops., the thoughts portrayed for David Barzilay and Century Books, by a Section Commander in 40 Commando who wrote down his views of the unit's two tours of duty in Belfast. He said his thoughts were so typical of what troops had to contend with during the act of "policing" civil unrest.

He served in the company group responsible for the New Lodge area during both tours. Gunner Robert Curtis, the first soldier to be killed in Northern Ireland during the early part of the campaign, was shot dead in the New Lodge in February 1971. During the second tour his company consisted of five rifle troops. With attached ranks they had about 170 men. Some troops were employed solely as patrol and O.P. troops in the centre of the New Lodge. The others patrolled the outskirts of the area, provided the vehicle patrols and reserves, and guarded the company base.

The foot patrol was the key to the problems of Northern Ireland. It dominated the area in which the terrorists operated. To be successful, a foot patrol depended on information. They obtained this from intelligence sources, from the local people and above all from the observation posts. The areas operated in were at their best modern flats and maisonettes; at their worst terraced streets which dated back over 150 years. His troop was split into two groups to carry out its task as an O.P./patrol troop. They were responsible for manning three O.P.'s on the high rise flats. One half of the troop manned the O.P.'s for 3 _ days at a time, while the other half patrolled the area around the O.P.'s for a similar period. Each time the O.P.'s changed, the same Marines went to the same position and took over the same watch. The reason for this was so that a Marine could get used to the routines of the day, such as the milk float on its rounds, the dustmen calling, the paper boy on his rounds and the shops and pubs opening and closing. In this way each Marine became familiar with the personalities and locality and was able to spot a change of routine when it occurred.

The O.P.'s were supplied when requested, usually during the quiet periods. Living conditions were very cramped. On the high rise flats they lived in the motor room housings for the lifts and shared the accommodation with two 1,000 gallon water tanks. The Royal Engineer Sappers did them a lot of favours in the way of equipment. The other half of the troop provided the foot patrols. They worked three hours on and six hours off for the same period - 3 days.

Each patrol consisted of an N.C.O. and five Marines. Anything larger was too cumbersome and too difficult to control. The O.P.'s tasked the patrols without reference to company headquarters. As soon as an observer spotted something suspicious he would task a foot patrol to the scene, pass all the relevant information on to it, and then observe for local reaction. At the same time a second observer would carry on with the normal observation tasks. This was a safety measure to try and spot if the patrol was being led into an ambush. For example, children causing aggro whilst a gunman lay in wait for a patrol. The O.P.'s were extremely useful during searches or riot situations. A section search team could be called back to their base location, briefed, and sent out again on a search mission within minutes.

It takes about three hours to search the average house thoroughly and this means a long boring wait for the Marines forming the cordon. It also allows ample time for a gunman to get into position and shoot at a Marine who is getting tired and losing his concentration. To combat this it was the Search Commander's responsibility to send out as much information as possible on the progress of the search in an attempt to hold everyone's interest.

A search operation often develops into a riot. This is particularly the case when the team has a find. Local sympathisers, normally women and children, will try every trick in the book to try and prevent the search continuing, and will attempt to rescue wanted persons and remove the evidence. Strong reserves are wanted in these situations, but even so large crowds of violent and abusive women and children are very difficult to handle at close quarters.

One of the tasks of the Search Team Commander was to keep a written record of everything that occurs. This was necessary to ensure that evidence was available for court cases and also to protect the searchers from any claims of improper behaviour. All members of the team would be searched before they started to ensure they did not "plant" incriminating materials and after the search to prevent accusation of theft.

Whenever there was a find the Military Police Finds Team would be called in to take photographs of it in situ and statements from the witnesses. Arrest - or lift - operations were never as easy as they sounded. When a Company Location Operations Room received information that a wanted person was in a house, speed was essential. A cordon would quickly be thrown around the house and the search party would gain entry and carry out a rapid search. Attics were always searched early on, because escape routes were often prepared in advance in terraced houses by "mouse-holing" through the walls.

Most wanted persons had an alias and documentary evidence to support it, so it often took time to establish the true identities of those found in a suspect house. Again, if any of the teams in 40 Commando were successful and were close to wanted persons, a crowd would form and attempt to gain access to the house involved, or endeavour to distract the Marines by creating a separate incident. The same people appeared time and time again on these occasions. One old lady who lived in a company location had been nicknamed The Witch because of the endless trouble she caused. Aged 72, she was a formidable opponent. They had to be watchful for every trick. On another occasion four men, blown out of their minds, approached the cordon and performed a drunken routine in an attempt to cause a diversion, only to be seen later, after the first volley of rubber bullets had been fired, to be the fastest, soberest runners in town. Another trick with the occupants of houses under suspicion in trying every which way they could to leave would be:

"I'm pregnant and I must get to the hospital";

"I've got a sick grandmother down the road and I must go and look after her".

The golden rule was – no-one leaves or enters until we have sorted it out. Often easier said than done!

While all this was going on on the ground, the O.P.'s would observe the neighbouring streets and keep Control informed of impending trouble. When the time came to remove any wanted persons the team would signal for an armoured vehicle or a closed Land Rover, bundle them into the back, and take them to Commando Headquarters so the arrest procedures could be completed.

When elements of the Commando were on foot patrol they moved tactically at all times, constantly on the alert for snipers, ambushes - rolling or otherwise - or a command-detonated explosive device. They always kept "one foot on the ground". Half the team stationary in good covering fire positions watching likely enemy ambush positions, the other half moving quickly and alertly to the next fire position. By day the gunmen stood a good chance, but at night the soldiers had a considerable advantage. With the excellent night viewing aid, or starlight scope, they could turn night into day. A hazy, greenish sort of day, but greatly effective. Some of this night vision equipment was first seen by the general public during scenes of the Gulf War on television. A section with a starlight scope was worth a troop without one. All patrols found that the opposition disliked intensely their soft shoe movements at night. As long as they knew where the Marines were, and in what strength, then they were happy. But just four Marines in loose order creeping around the streets threw them into fits of confusion. If accidents were to be avoided, however, it did mean that O.P.'s and other patrols had to be thoroughly briefed.

Humour played a big part in Northern Ireland. Tight discipline with good humour sometimes had more effect than a whole volley of rubber bullets. There were a great many people, even in the New Lodge Road, who were pleased to see the Marines patrolling the street. Many people gave words of encouragement as the teams passed through. Sometimes a whisper and useful information as they passed a doorway; or at night a quick wave and a smile. This was also where the intelligence factor came into it, and also why a patrol never ended at the main gate. They would grab a quick mug of tea, and in a businesslike atmosphere the patrol would discuss and piece together every bit of relevant information, write it down and pass it on to the company int. section.

The most important man in Northern Ireland in those early faltering days was the Marine or other soldier on patrol. It made no difference whether he was operating against the gunman or the rioter, he had the chance to influence events for the good or bad. Most soldiers were completely impartial and didn't mind who they were operating for or against. Every man in a section had his own peculiarities and skills and any Section Commander worth his salt should have been able to spot them and help develop them. If the leadership was good, this would reflect on the streets. The Marines had the best. As a result, the foot patrols, O.P.'s, mobile patrols and other elements were able to give of their best with the certain knowledge that they would be backed up to the hilt.

Most of what we have just been talking about were the thoughts, as I mentioned before, of a 40 Commando Section Commander who spoke to David Barzilay, the author of The British Army in Ulster (Vol. 2) – Century.

CHAPTER 20

45 COMMANDO GROUP: JULY 2ND TO NOVEMBER 5TH 1974

45 Commando Group carried out a 4-month emergency tour in Northern Ireland between July 2nd and November 5th 1974. When told the unit would be responsible for Police Division "H", later termed "Bandit Country", I knew, even before we left "Condor", the unit location in Arbroath, the chances of completing a 4-month deployment around those border regions without at least some injuries were remote. We were going to the Provisional I.R.A. heartland, their training areas, their constituency, their home turf, or - if you like - their virtually unhindered stomping ground. In this theatre of operation most of the basic infantry skills were utilised, and junior leaders down to section level went through many valuable field experiences. The three rifle companies in the Commando "X"-"Y"-"Z", X-ray, Yankee, Zulu, had areas which varied considerably and each required an entirely different way of operating due to the terrain and the differing tactics employed by the terrorists. The first hurdle to be overcome was that of the intelligence factor, or "getting to know" the ground, the people, and most important of all the terrorists or "players" as I always call them. It took several weeks for the rifle companies to become acclimatised to the locals and their often "strange" ways. But soon, armed with this knowledge, the Commando successfully carried out offensive operations in the border areas of South Armagh.

Throughout the whole of this tour I was with "X"-ray Company at Bessbrook Mill which is very close to the Irish border, and in 1974 a hotbed of P.I.R.A. activity. I also had the freedom and flexibility of conducting ops. of an int. nature in civilian clothes which had the drawback of me being dumped at a moment's notice, at all hours of the day and night, in all weathers and local political climates, into the most hair-raising situations imaginable. On one or two occasions, for reasons I cannot go into, I can say with certainty the only way to avoid the attentions of the Northern Command I.R.A. Internal Investigation Units, which feed on int. from the locals, was to produce a civilian tourist map and revert to my "Terry Wogan" accent as a visiting Dubliner lost in South Armagh in November. Not very convincing on a wet, windy day, but somehow it always worked for me and I made it back to Bessbrook every time.

To give a broad and typical outline of a Royal Marine tour of duty in Northern Ireland I will use this tour, the most interesting of my six, to bring home the problems of operating in such an environment as so well outlined in the coverage by David Barzilay in "The British Army in Ulster" (Vol. 2), but I will inject into that coverage many incidentals from my experience of having been in situ, as it were.

Let us now start with an incident on the very day of our takeover on 2nd July 1974, when a proxy bomb in a cattle truck was driven to Markethill where the R.U.C. constable at the town barrier forced the driver at gunpoint to drive the vehicle out of town. As a result, the 400 lb bomb exploded harmlessly away from buildings and people. Intelligence was quick to find out where the bomb had been loaded, but this led to "Felix" (the Bomb Squad) searching a derelict old farmhouse, which resulted in the death of Royal Engineer Sapper Walton. This, no doubt, was just what the terrorists had hoped for, and this trap was in fact set in the hope of catching some of the newly-posted Marines. The type of device that was used to initiate the explosion was never discovered, but it demonstrated to all of us in "X" company, and especially to me in intelligence, just how determined these terrorists could be. I then immediately but reluctantly had an enlarged photograph produced of the death scene of Sapper Walton, which included part of his clothing seen through the rubble, and posted it on the "Watchword" notice board at "X" Company location at Bessbrook Mill. Yes, it raised a few eyebrows, but more to the point it raised many more senses as to the immediate type of enemy at our gates. That was my point, and looking back, I think it worked.

TRAP SET FOR ,MARINES

On July 11th 1974 the Marines suffered their first casualty in Crossmaglen (X.M.G.), when Marine Rennie was injured by an explosion in the town square. The device, about 5 lb of explosive, was detonated by radio control as a foot patrol passed the town hall. Marine Rennie's flak jacket and rifle took the brunt of the explosion and his injuries were not severe. Later that day an officer from bomb disposal burned the remainder of the already derelict town hall to the ground to make very sure it contained no more explosives.

One of the more sensitive disputes that the Marines had to contend with during this tour was the dispute over the control of street lighting in the centre of Newry. "Z", Zulu Company, based at the Ulster Defence Regiment Centre, Newry, could, if necessary, turn out the street lights in the town from one central switch, the advantage of this being any patrol which came under fire at night could call for the lights to be switched off, which would give the patrol the cover of darkness. This also avoided the shooting out of street lights which was both wasteful, and involved the use of weapons, which could be so easily misinterpreted by the local population.

Problems with the electricity supply in Newry started on Wednesday August 7th, when maintenance workers employed by the N.I. Electricity Service refused to carry out repairs to equipment because of threats from the Provisional I.R.A. These threats, the P.I.R.A. said, would continue until the Security Forces returned control of the street lights to the Electricity Service. Gradually more and more breakdowns occurred and by August 22nd practically the whole of Newry was in darkness. Attempts were made by the Security Forces to provide generators so that essential services could be maintained. The local people refused to accept this assistance, again because of P.I.R.A. intimidation. The P.I.R.A. had indeed come up with a new and effective weapon, but the people of Newry remained calm despite the inconveniences they suffered with no power supplies.

Attempts to break the deadlock were numerous, and included handing over control of the lights to the R.U.C. But the locals did not trust the R.U.C. Negotiations between the Marines, the local politicians and the N.I. Electricity Board went on but failed to come up with any sensible solutions. The crisis eventually came to a very sudden end on September 4th, when Mr. Stanley Orme, Minister of State for N.I., announced that the Army would hand back control of the lights to the Electricity Board. No indication of this decision was given to 45 Commando and on September 5th the Loyalists News carried headlines such as "Anger over P.I.R.A. Victory", "Rees gave in to the Provos", and "Loyalist Rage Greets Newry 'Switch On'."

The month of August will be remembered not only for the Newry lights, but also for attempts by the I.R.A. to disrupt and block the Belfast-Dublin railway line. A train was hijacked on August 3rd just north of the border at 04.40 hours. The train, consisting of a locomotive and one coach, was crossing the border when the driver saw a flashing red light on the track ahead. Just as the train slowed to a halt, six or seven masked men emerged from the darkness, jumped aboard and ordered the driver off. The terrorists then loaded "packages" into the engine compartment, and escaped over the border south. Once the position of the hijacked train had been established by air recce, a major clearance operation was mounted and the Chief Bomb Disposal Officer in Northern Ireland at that time, Lt. Col. Gaff, tasked his team with the job of neutralising the devices. Elements of 45 Commando staked out the ground and prepared themselves for a stay of up to 48 hours for this operation. The train was not in an ideal position for the explosives team to start work on neutralising the packages, so an attempt was made by N.I. Railways to push the hijacked train further north to a disused station. But as with a lot of situations like this, good old "Murphy's Law" took over again and moving the train failed, so bomb disposal made their way along the track and eventually defused the packages, which were found to contain collectively approximately 400 lbs of explosive. This operation was completed by 21.30 hours. The Marines were on the ground for just 17 hours from the time of hijack. Seven days later yet another attempt was made to block the line when an aqueduct one mile north of the border carrying a stream over the railway, was blown up. Damage to the line was minimal, and rail services only slightly disrupted. On August 21st the early morning train from Dublin to Belfast was again hijacked, this time at 05.20 hours, and a mile further north of the border. The Commando mounted a major clearance operation and Major

TRAP SET FOR ,MARINES

45 Commando Royal Marines regrets the necessity for delaying your journey and wishes you a safe arrival at your destination.

SUPPORT THE COMMANDOS
AND
HELP US PREVENT VIOLENCE

PLEASE
REPORT ANYTHING SUSPICIOUS
AT ONCE BY USING
The Confidential Telephone
NEWRY 3015

INFORMATION MAY BE GIVEN IN
COMPLETE CONFIDENCE

Pickard, the Chief Bomb Disposal Officer's deputy, took over the unenviable job of defusing the device. The initial approach was made in the same way as his Chief had done on August 3rd, except that Major Pickard put a dummy on the train instead of himself. When eventually he made an entry, a 100 lb device exploded at the opposite end of the train to which he had entered.

A third hijack of the morning train was made on November 5th, just eight hours before the Commando handed over the area. Fortunately the terrorists had made an error and the train actually stopped in the Republic of Ireland, just 100 metres south of the border. This time the Commando was able to relax and watch how they did it on the other side.

During my tour of duty in "bandit country" with "X" Company, the Dublin to Belfast railway had suffered considerably during the troubles, and it was a very easy target for the terrorists. Lengthy operations were mounted by various elements of the Commando in an attempt to trap the hijackers, but without success. The disarming of the devices in August, and the subsequent saving of the engines gave the morale of the local people a much-needed boost. The terrorists were seen to be defeated in their attempts to kill bomb disposal officers, which seemed to be the primary aim of the hijacks. But it did not end when we left the Province on the 5th November to return home to 45 Commando base in Arbroath. Later we heard of many more heavy bombings of the Belfast to Dublin railways.

"You Can Break... But You Can't Quit"

On August 13th the Commando had their first and, as it turned out, their only fatal casualties during this tour. Corporal Leach and Marine Southern were both killed by a 100 lb bomb at Drummuckaval two miles south east of Crossmaglen and 300 metres from the border. Very early that morning at Bessbrook Mill I had been woken up as part of the intelligence team to prepare some maps for "X" Company, who were making preparations to visit and search several farms on the border regions. I was alone in the galley (cookhouse) where I had joined two tables together to spread out, mark up, cut and waterproof the maps which were needed by the troop and section commanders of "X" Company. As I worked, the galley door opened and Micky Southern walked in. I don't know how long he had been at Bessbrook that morning or what his last duties had been, but as a long-time friend of his I could see that he was depressed and worried about something. At the time I didn't put this down to much, because most people working in this environment were not always at their best. But, as well as I knew Michael Southern, I knew – and somehow I think he knew – that things could go wrong at Drummuckaval. I don't know where the rest of his team were, but he just filled a black plastic water bottle mug with tea from an urn and, as he walked back out the door, he simply raised his free hand and smiled at me across the tables. The door slammed shut and I never saw him again. I know absolutely nothing about premonitions or such things, but the man who went through that door was not the ever-grinning optimistic fellow I had known for so long.

Corporal Leach had been manning the Drummuckaval observation post for two hours, when at 15.00 hours Micky Southern came to relieve him. Just as he entered the O.P., the 100 lb device, which was radio controlled from across the border, was detonated and both men were killed instantaneously and two others were wounded. The sheer destructive power of 100 lbs of explosive at close quarters can be realised by looking at the remains of the Marines' weapons, which I photographed after the event. The body of the general-purpose machine gun was torn apart. The heavy barrel was bent like plastic. On the linked belts of ammunition the bullet-heads had been torn from their brass casings and found up to 100 metres around the site. The S.L.R. rifles were reduced to twisted steel; most of their woodwork was gone and never found. The smell of burnt earth and explosive residue was overpowering. I spent as little time there as I could, but the aftermath of cleaning up and investigating had to be done and later that evening, as glad as I was to leave that place and as sick as I was at the sight of everything, I know Corporal Leach and Micky Southern would have been the first to remind me of the sayings of so many instructors in basic training: "You can break, but you can't quit." Amen to that.

How long the bomb had been in the O.P. will always be a mystery, but the Marines were again reminded that they were fighting a determined and sophisticated enemy, an enemy who were

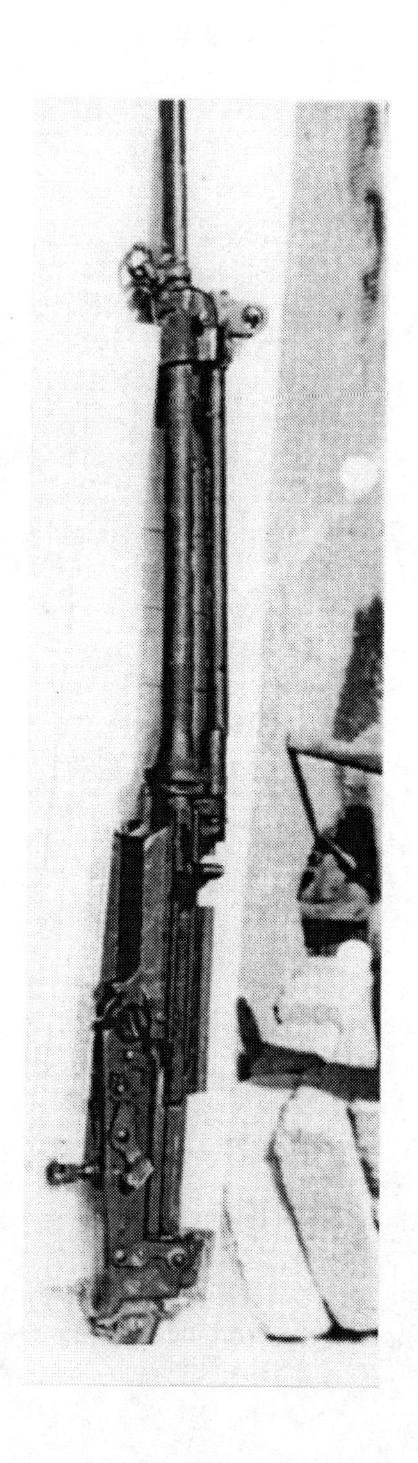

SLR RIFLE TORN APART

REMAINS OF LINKED AMMO BELT

REMAINS OF MACHINE GUN G.P.M.G

REMAINS OF OBSERVATION POST

prepared to sit and wait for the right moment to strike. After this incident the Marines built up a permanent overt O.P. at Drummuckaval from August 13th, with the aim of preventing the terrorists getting into Crossmaglen to positions where they could attack the R.U.C. station.

During the tour the Commando succeeded in this aim, and only suffered one attack in the vicinity of the R.U.C. station, which occurred as a patrol was leaving. One man, Corporal Buchanan, was hit five times, but did not suffer serious injury. The O.P. was in fact attacked by mortar and small arms fire on eight occasions, but no casualties were sustained. One home-made mortar bomb actually fell on the O.P. and exploded harmlessly when it was being manned over a weekend by men from (3) Ulster Defence Regiment.

On the afternoon of August 18th, news reached the Commando of the escape of 19 I..R.A. men from Portlaoise Prison in the Republic of Ireland. Border operations were immediately stepped up, but were hampered by the lack of up to date photographs of the escapees. The very prospect of having an additional 19 I.R.A. men in the area was depressing, to say the least. On Friday September 13th, "Yankee" Company managed a spectacular success when a foot patrol spotted three gunmen moving on the outskirts of the town. The men were challenged and the gunmen turned to run. The patrol opened fire, seriously wounding one of the men, but the other two escaped. The identity of the gunman was not known initially, but it was soon discovered when the family of the man, Martin MacAllister, a Portlaoise escapee, rang the Commando ops. room asking which hospital he had been taken to.

During the tour the Commando fatally wounded two people. The first was on August 4th when a patrol of 1 D.E.R.R. operating just outside Castlewellan shot and killed a man just as darkness was falling. The P.I.R.A. claimed he was the Adjutant of their South Down Brigade. He was armed with a .303 rifle at the time. The second success was in Newry on October 18th when an O.P. shot and killed a 16-year-old, who the Army alleged was in the process of carrying out an armed hijack. There had been a spate of hijacking in protest against the Maze Prison riots in Newry for the 24 hours before the shooting. "Zulu" Company subsequently sent an O.P. to watch the Derrybeg Estate, an area in which hijacking inevitably started.

The O.P.'s vigilance was rewarded when a local bus appeared from the estate with an armed man forcing the driver to position the bus across the main road, with the intention of eventually setting it on fire. When the bus reached the main road, the gunman got out and was challenged by the O.P; he turned towards them as if to fire and was then shot by a single round by one of the O.P.

Reaction to this shooting was very much quieter than expected, and low level intelligence which I later read at Bessbrook indicated that the people of Newry were not at all sorry that this man had been shot.

Moving on towards the end of the Commando's tour, plans were made to start closing the border crossings and thus making the terrorists' getaway more difficult. There were 43 crossings altogether in the unit's area which gave us a variety to choose from. The Commando had learnt an early lesson about border closures in the first weeks of our tour near Omeath. On June 28th, 7 Para. blocked the road, but within 24 hours the locals had filled in a dried up stream and by-passed the obstacle. The Paras. arrested varying numbers of people trying to cross and the block was improved. On the night of the 4th/5th July, by which time the Commando were in charge of the area, the obstacle was removed completely by plant and acetylene cutting equipment. The Royal Engineers replaced the obstacle, and an O.P. was set up close to the block to ensure it was not removed again.

These incidents were gaining considerable publicity at this stage and the Commando Public Relations Officer was getting his first taste of dealing with the media. Eventually political negotiations with the Dublin Government were successful and they set up an Irish Army post close to the crossing, and the obstacle was removed. Later the Commando closed several other crossing points. Co-operation with the Gardai did improve during the tour, and so helped with this and many other problems.

Proxy bombs were a continual source of concern, and there was much discussion as to what action the Marines should take when on the receiving end of one. In my area, Newtownhamilton was a prime target and on September 4th an ice cream van was hijacked between Newry and Newtownhamilton, a bomb placed in the back,

ICE CREAM VAN BOMB

and the driver ordered to park the vehicle as close to the R.U.C. station as possible, which he did. It was not possible to get a "Felix" or any other bomb experts to the scene within about half an hour, so clearance was given to fire a Carl Gustav anti-tank round at the vehicle in an attempt to dislodge the device. If you can picture this scene, and for the benefit of readers who may not know, a Carl Gustav is an anti-tank rocket-firing shoulder held weapon which the Americans call, or like to call, a Bazooka. Well, if they dubbed me the "loose cannon", the man they chose to fire this weapon was indeed a true and trusted "wild card". Because I haven't been able to contact him for permission to use his name, let's just call him "Yorky". After some rapid preparation and planning, he got himself into a good firing position and without further ado he released the 84 mm round straight at the ice cream van. Unfortunately, this in turn set off the 400-600 lb bomb causing extensive damage to surrounding property, but no loss of life. It also spread huge quantities of ice cream over the residents of the town, some of whom later made massively exaggerated claims for destroyed clothing on washing lines. When some of the claims came in, it seemed obvious that this was the year when everyone in Newtownhamilton was wearing Gucci or Giorgio Armani, and to cap it all, they were compensated.

The following day it was also reported that cows in a field up to 125 metres from the point of explosion were still contentedly slurping the now melting ice cream, which was leaking from dozens of fractured plastic containers. The farmer seemed to agree that any trauma the cows may have suffered from the big bang they heard was quickly compensated for by these strange gifts which fell that fine sunny day from the clear blue skies of Armagh. So, here yet again was a perfect example of a situation where a sense of humour injected into a God-awfully dangerous event did more to ease the problems of civil unrest than rushing around en masse, kicking doors down and searching for a proxy bomber, who was by then more likely to be in Dublin than Armagh.

Another proxy car bomb was driven to the Omeath vehicle checkpoint from the South on September 6th. The Marines on duty at the time apprehended the driver, who tried to run away, "persuaded" him to get back into the car and drive it 400 metres up the road, which he duly did. The car exploded 15 minutes later causing no damage. Another good result.

For the terrorists, two "own goals" occurred in my area during the tour of duty, one in Newry, and the other on the main Dublin road south of Newry. In the first incident on August 28th, one man, Patrick McKeown, was killed and a small girl injured when a 20-50 lb bomb exploded in a house. An obituary notice said he had been "killed in action".

The second incident occurred on October 5th when a Mr. Eugene McQuaid blew himself up whilst riding a motor-cycle on the Newry-Dundalk road. Investigations revealed that he was carrying home-made rockets, one of which had exploded.

On Saturday October 19th the Commando carried out an operation which later became known as "Operation Kettle". On October 17th, a vehicle which contained £15,000 of electrical goods had been hijacked close to the border village of Jonesborough. The R.U.C. received information that many of the village inhabitants had been given the equipment, with a view to it being sold on the "open market" held in the village every Sunday. Support helicopters were tasked to assist in the operation and a company landed and cordoned the area. Visits to houses soon revealed what they were looking for, and electric kettles and fires, fans and toasters appeared from everywhere. The local village hall was taken over to accommodate the villagers who had been arrested. Eventually 25 people were flown to Bessbrook Mill to help the police with their enquiries.

During their tour the Commando experienced the anniversary of internment, on August 9th 1974, which saw large scale hijacking in Newry, and a U.K. General Election, which passed off without major incident. We summed up our tour by saying: "If one had to be anywhere in Ireland, South Armagh was the place to be. The area and the people provided a tremendous challenge to any unit deployed there."

Earlier in this book I promised I would not overburden you with any one particular theatre of operation for the Royal Marines, so let us stick by this with just one more story which needs to be told if for no other reason than to prove the madness in men and the risks they take are truly infinite.

“The Trick... Was to Stay Alive”

It was just after 2 o’clock in the afternoon on October 12th and it was dry, hot and airless in the operations room. My desk was piled high with mug-shots and personal check cards of the dead, the imprisoned and the wanted. Both Alan Jones and Jim McNeill, my contemporaries on the intelligence team, were at their desks just opposite me. I stood up, pushed the big side window open, but it was still stifling. A small wall hatch next to Alan Jones’ desk flew open where the Company Commander just about managed to fit his head and shoulders to warn us an amount of explosives had appeared in a disused farm building on our patch quite close to the Mount Norris area. The warning had come in anonymously, for many of these people are too frightened to give their names. So, here we were in the ops. room trying to decide between us if this was basic good intelligence from a law-abiding citizen, or was it a come-on, a trick to draw us to the scene where we would be met by a well-placed ambush of two or three “players”, with assault rifles of the finest calibre which were now finding their way into Northern Ireland from all over the world.

A decision was made. We would take an old Q. van, an anonymous transit from a pool of old vehicles which we kept for just such purposes, and let the driver drop the three of us at a point some two to three hundred yards from the farm and then make our way in tactically, with the use of what cover we could find. We were in civilian clothes, we each had a 9 mm Browning auto pistol, but Jim and I also carried S.L.R. rifles with spare magazines stuffed in our waist belts. Not very original I must agree, but let’s face it, in those early days in Northern Ireland when nothing was working, not even the policy of Government, we really were playing some very “wild card” games over there. And they called me “The Loose Cannon”!

At the chosen drop-off point where the driver would then wait, Alan, Jim and I made our way to the target farm without incident. The old building was desolate and bare except for some old ragged curtains on the windows. It was open land all around and the neighbouring farm animals seemed to come and go as they pleased. At the north-east corner of the building across a laneway stood a big old chestnut tree on a raised mound which gave us a perfect area of high ground from which to cover Alan while he made his way forward to a corner window to look inside.

He could clearly see in the far corner of a front room some industrial fertiliser bags, which was the material used for bomb making and very easy to come by in the early 70's when such devices were causing havoc in most of the towns and cities. Alan moved back to our position for a quick briefing. The situation suggested that this material was just awaiting collection, which could be in two hours, or it could be two weeks. This called for one of two things; an O.P. to watch and record what happened and perhaps follow up on it, or a blatant ambush position to allow the collection to go ahead and then destroy the players right there on site. Well, these moral issues were certainly not for us to decide on this particular occasion, so our decision was for Alan Jones to go back with the driver to Bessbrook Mill and allow the Company Commander to make a decision on sending an ambush group out to replace Jim McNeill and myself, as we were now alone on our overgrown mound by the chestnut tree awaiting events. What hapless politician once remarked that his downfall could quite easily be brought about by: "Events, dear boy, events!"

Well, it's true that events can kill you and I knew we were right out on a limb that afternoon, and now with Alan and the Q. van driver gone, we were totally exposed to any "events" that may occur at this bombers' lair.

It was mid-afternoon, dull and overcast, calm and humid for October. The only sound to split the silence on the odd occasion was the hunting of vermin on the surrounding farmland and wooded areas by the local farmers. They seemed to have an obsession with this and legally licensed weapons were common and frequently used. This, needless to say, did nothing to reassure two lone men in a hastily set up observation post with no idea in what strength the enemy might turn up to collect "their" ordnance from "their" bomb-making factory.

In the first fifteen minutes of our watch we had settled down, about five yards apart but low and flat amongst the grass and weeds with a good view of the old house, within a distance of about thirty metres and an angle of 5 o'clock. Between us and the house ran the narrow gravel laneway. During that first fifteen minutes three distant shots rang out. We had no option but to hope they were just hunters on their kills and nothing more sinister closing in from various angles. We had, of course, no way of knowing. We could only hope the Company Commander would quickly read the urgency of the situation and in all haste deploy an ambush team from 'X' Company. And then it started.

The best-laid plans of mice and men, call it what you will. Murphy's Law, events, dear boy, events, it doesn't really matter now, but events overtook us and the plan went out the window. Jim McNeill was on my right looking half-left at the approach part of the laneway. I, after spending several minutes flicking industrious ants from my arms and neck was now resting my chin on the butt of my rifle which I had in an upright position supported by its magazine and pointing at the garden of the house. I think I heard it before Jim did and as I glanced at him, his expression changed to a recognition that something was wrong. The shuffle of feet on loose gravel to my left on the lane was unmistakable. I glanced at my watch, which gave me 17.30, and then back again to the lane approach and a truly amazing sight. A little old man in a duffle coat of sorts, which was ridiculously too long for him, bumbled along straight towards the little garden entrance to the house. In he went through the front door and closed it after him. The house was in such a state of dilapidation I knew he wasn't in residence, so we waited. Five minutes later he came out and pulled the door closed behind him. He was empty handed and he just stood in the garden and looked all around, his beady eyes calculating and penetrating. At this point I think Jim and I must have sunk a further three inches into the earth just willing him not to see us and it seemed to work as he just cast his glance through our position and away.

Between Jim and myself there was a silent conversation going on in as much as we now knew that this man was conducting what was in all probability a sweeping exercise to clear the building of intruders in advance of the main body of bombers coming to pick up their materials. Or was he? If we let him go, we might lose him and his knowledge of this house forever. Well, there was no time to wait for a House of Commons vote on the infringement of his human rights which seems, as the years roll by, to become more and more "politically correct" nowadays, almost to the point of writing a letter to a terrorist's solicitor warning him that his client might get shot by the security forces if he continues to plant bombs in fish and chip shops on busy Friday evenings. So, Jim and I knew what had to be done and without a spoken word between us he covered me while I slid down and around my hillock and like a ghost in the night hauled this latter day "Tonto" by one arm and the neck of his duffle coat back and around to our position. I planted him down next to Jim where we both read him the riot act after searching him, warned him that if he did not sink himself down into the weeds and spend the next hour or so of his life in utter and complete silence we would cut him into a thousand pieces and "feed him to the Gurkha's in Hong Kong". This had the desired effect, but I was later convinced that this poor old yokel didn't even know where Belfast was, let alone Hong Kong.

Now it was back to the waiting game, the watching and wondering. It lasted just 35 minutes before the crunch of tyres on the gravel laneway. There was no caution on the bombers' agenda whatsoever, no suspicion to encourage even the slightest stealth into their approach. Suddenly they were just there. A white Ford Anglia in front with the two main players, and behind them a yellow transit van with four foot soldiers.

"Assessment of the Situation"

When the six of these men left their vehicles and gathered in the garden of the house by the doorway, the first sign of a difficult situation – to me anyway – was the lack of any visible small arms. Armed terrorists or paramilitaries can be shot if they show disregard or aggressive opposition to any lawful directive from the security forces. This is not something which was drilled into me by some Sergeant-Major, but common sense which runs directly in line with my way of thinking. I will, no doubt, be berated for my views but I will give them anyway. I have said this in "The Fourth Province" so let me run it past you one more time. Understand something – and it's really so basic: when it's my life on the line and when I'm there in the killing zone in a combat situation, there are no holds barred, O.K! Anything goes, everything goes. You know, it just comes to you that you want whatever is possible at hand to save your life and destroy an enemy who is trying to kill you.

THE MOUNT NORRIS BOMBERS LAIR

THE MOUNT NORRIS BOMBERS LAIR

BOMB MAKING KITS & OTHER ANCILLARIES IN THE HOUSE

I would give no thought whatsoever to his human rights, minimum force, political status, and I certainly would not entertain even the farthest margins of political correctness. I would want a rifle to inflict as much damage as possible, without any regard to the pain and the suffering that I might inflict on an enemy. I would want a rifle, I would want a bullet, I would want any weapon, I would want anything that is going to kill an enemy who is trying to kill me or the people around me who are dependent on my performance in the field. It's as basic as that. So, here lay Jim McNeill and myself, the village idiot "Tonto", and six apparently unarmed terrorists.

If the latter were true and the bombers were in possession of no small-arms at all, then we really did have a problem here. As I read the situation, three things could happen now at 18.15 on that evening of 12th October. If the six players, after we confronted them, came to realise we were just two members of the security forces and they indeed were in possession of no small-arms at all they could, if they were a well-drilled and experienced team, simply hurry back to their vehicles and drive away. And what could we do about it? Even though we had caught them in the act of removing and loading the bomb-making materials onto the yellow transit, if Jim and I slipped our safety catches and shot them all down we would risk a minimum of ten years in prison. That was their first option. Their second option, if they had the required guts and determination, would be to split and spread out then come back at us from six different directions and once again, without actually shooting them, who can say what might have happened, bearing in mind had they found a way to overcome us numerically, not only would they escape with their bombing kits, but also two S.L.R. rifles and two 9 mm Browning Autos. And if this did happen, it would have happened because we were too shackled and enmeshed in political correctness to call a spade a spade and just get on with it and do what had to be done. What, me a loose cannon? Never.

Now, the third way to my reading of this "fun day in the country" was the one sure way to proceed, now that I had challenged the six players and had them face down on the grass in the small front garden. An idea was slowly taking shape in my mind. On the side of the old yellow transit van in dirty and fading letters was something in relation to Harland and Wolff Shipyard. Knowing that most of the troubles of Northern Ireland stemmed from the problems of civil rights, the shipyard was a perfect example of the rot that bedevilled the Province on a daily basis. There was a point when it was correct to estimate the working population of the shipyard was 97 per cent Protestant and about, let's say, the remainder Catholic and others. Such was the inequality which festered and bred the ongoing hatred.

In view of the markings on the van it would be reasonable to calculate that the six players were Protestant Paramilitaries. I could have been wrong, but it was all I had to play with within these short few minutes of calculated mind games. As they lay flat on the ground with their faces on the grass, I knew the next paramount move was to dispel any notion they may have formed in their heads that they were being held by just two members of the security forces. This was essential before any one member of the six decided to encourage the others to collectively mount a challenge. These Paramilitaries were well versed with the law and they knew if we were military we could not shoot them.

The answer that came to me was to save the day. It had thrown me a lifeline in Republican Ballymurphy just as it had caused untold alarm in R.U.C. canteens. The players needed frightening into total immobility. I quickly moved to my right until I was next to Jim where I gave him a fifteen second briefing on the plan. I left my rifle with him, pushed "Tonto's" face down into the weeds again and warned him one sound and he would be on a plane for Hong Kong and the Gurkha Brigade cookhouse. He was very good really, but also very confused. On moving back to my original position I took the 9 mm Browning from a shoulder holster which was out of view under a light windproof jacket, re-checked its firing status and its 13 round magazine. I had 2 more loaded mags in my pocket. Once you come to know and control the 9 mm Belgian Browning Auto, for C.Q.B. in confined spaces you have a concealable killing machine. Other than recreational target shooting, the Browning has no other forward purpose than to blot out life. This was all I needed for my next move. Within seconds I was down the side of the hillock across the gravelled laneway, into the small garden and over the six players. From left to right I searched each in turn for weapons but there were none. As I searched I allowed the cold slide on the side of the Browning to rest against the face of each individual. They were experienced players and anything but stupid. They could smell the unmistakable combination of oil and burnt cordite from the weapon, which I fired every other day at the tunnel range in Bessbrook Mill, but the real fear which seemed to overpower them now was their captor berating them in an accent which was straight from the streets of Dublin. It threw them completely and this confirmed my assumption they were Protestant Paramilitaries. They confided to me on a much later date that on 12th October at the time of their capture they were absolutely convinced they were being held by Provo's from the I.R.A.'s Southern Command of Dublin, Cork or Kerry.

On an overall assessment of this situation it could have turned from nasty to catastrophic. On seeing how outnumbered we were, we could have made a tactical retreat for the sake of saving life and limb on both our side and the side of the terrorists. But the easy way out is not always the practical way, remember you can 'break' but you can't 'quit'. As fate would have it on that particular day for both Jim McNeill and myself, let us just say that fortune favoured the brave and we won through without having to kill or be killed.

As it turned out we held the players for another half hour before Alan Jones arrived back with an ambush party (call sign one/two), who then took charge of the whole show. Jim and I handed a very brave but very confused "Tonto" over to call sign one/two, who it turned out was later released with no charges. Why? Don't ask. The six players were sentenced to eight years imprisonment for plotting to cause explosions.

Later, the General Officer commanding Northern Ireland awarded both Jim and I commendations for services rendered. I have to say the following year I was not even "nominated" in Hollywood for an Oscar for my performance as a mad Irish terrorist. Quite disappointing really, but one thing was for sure, I certainly fooled those six players. I could tell you many more stories such as this involving 45 Commando, but I promised I would keep it short, so we will now back off the subject of Ireland with these last few thoughts.

Lieutenant-General P. J. F. Whiteley, OBE
Commandant General Royal Marines
Ministry of Defence
Main Building
Whitehall
London S.W.1

01-218-7675

RM 17/1/73

Marine D J Griffin
41 Commando Group
St Andrews Barracks
Malta
BFPO 51

16th August 1975

Dear Marine Griffin,

It is with the greatest of pleasure that I write to congratulate you on being awarded the GOC's Commendation for service in Northern Ireland.

Your performance in 45 Commando Group during the Unit's last Northern Ireland tour reflect much credit upon yourself and the Corps. Well done and best wishes for the future.

Yours sincerely

Peter Whiteley

From Lieutenant General
Sir Frank King KCB MBE

HEADQUARTERS
NORTHERN IRELAND
LISBURN
CO. ANTRIM

Lisburn 5111 ext 400

20 July 1975

Marine D J Griffin RM
45 Commando Royal Marines
Condor Barracks
Arboath
ANGUS Scotland

Dear Mne Griffin

I am writing to say how pleased I am to award you my Commendation for your service in Northern Ireland.

I know that on the 12th October last year how well you reacted during a situation when six terrorists were arrested in the Mount Norris area. I am delighted that your conduct at this time has received official recognition.

Well done, and every good wish for the future.

Yours sincerely

Frank King

Headquarters Northern Ireland

MARINE DAVID JOSEPH GRIFFIN

ROYAL MARINES

Your name has been brought to my notice.

I am authorised to signify by the award to you of this Certificate my appreciation of the good service which you have rendered.

I have given instructions that a note of your devotion to duty shall be made on your Record of Service.

Frank King

Lieutenant General
General Officer Commanding
and Director of Operations
Northern Ireland.

Date 20 July 1975

Over the past 35 years so many articles and books have been written on "The Troubles" of Northern Ireland that quite apart from the unfortunate people of both our nations, not to mention so many of the publishing houses of the English-speaking world, people in general are sick to the back teeth of the persistent back tracking, its inconsistencies and double standards. To me, Northern Ireland is one of those places where it would be considered wise to believe only half of what you hear and believe at face value nothing you see. The streets have born and bred a whole world of actors and illusionists where the age old "bull" piles up so high all around you, one would need gossamer wings just to stay above it all. The wild horses are held in check now by a few moderates, such as David Trimble (at time of writing!), who clings to some half-hearted hope of deliverance much akin to the antelope in the jaws of the lioness. In the shadowy sidelines lurks a whole new generation of combatants just waiting for the changing political climate to establish their target definition.

When this happens, and it will, any steady influences we now enjoy such as Trimble, and yes – believe it or not – McGuinness and Jerry Adams, will be swept away like so much old and unwanted dust from a picture frame. From the shadows will emerge the new warriors with their new excuses, their new weapons, new political points of view, new programmes of mayhem and organised calamity, their frightening new streetwise strategies in the art of planning at all levels will roll out to be tried and tested.

"Last night two more soldiers…", the dreadful weary words on most news bulletins throughout these early years of a new century. During the never-ending learning curve against terrorist activity, the long lonely road of experience to a new kind of war which continues to confound successive governments and security agencies and if anything is growing at an alarming rate continues to build in sophistication and methodology. The problem seems to be, you do not defeat terrorist actions, you mould them into what you can reasonably term an acceptable level of violence. Then you had better learn to live with it because you will never completely discipline or control it any more than you will with drink, drugs or illegal weapons.

Britain learned this hard lesson from years of trial and error in Northern Ireland, especially in the very early days of internment, which turned out to be the greatest campaign for I.R.A. recruitment in 80 years. Volunteers ran in droves to the Provo's and as a result of this, home-grown terrorism had to be lived with and bedded down.

Most people who know and study the problems of terror or paramilitarism will by now know that both the British and Irish Governments are fully aware of massive arms dumps and their precise locations all over Ireland. But they will tell you that inaction on such armouries is the preferred method of containment. Each and every government for over 30 years now, have served their time ensconced in the fairy tale adage: "Leave them alone and they will go home dragging their 'arms' behind them." They fear moving in on the terrorists because they know an explosion of violence and retaliation would follow from both sides of the divide. Even in the south, or should I say especially in the south, the Dublin Government and what security forces they could manage to cobble together, would be at a complete loss as to how to handle the problem.

For many years now consecutive Prime Ministers, both British and Irish, have said a silent prayer every day that terrorist players of every influence would continue to hold fire until their terms in office were over, when they could dump the whole theatre and the related top secret files on to the next sphere of responsibility. This never-ending game is played out because, while we can wage huge retaliation on to desert dictators who harbour terrorists, our own home-grown paramilitarism leaves the democratic system slavishly loyal to domestic human rights and the manipulative mind-games of the camp-following lawyers. The result is frightening impotence.

Most terrorists are enveloped by a steel wall of protection from the legal system, the much overplayed "Human Rights" card and, dare I say, democracy itself. It was none other than Winston Churchill himself who in the office of P.M. said in essence, "Democracy is a dangerously awful way to live, but until we come up with something better, we're stuck with it." He knew how the system allowed the bandit to slip away laughing and it goes on today. I have always had a good idea of who, across the border, was instrumental in the deaths of Corporal Leach and Michael Southern as there are those today who know the whereabouts of the Omagh bombers, but direct action would not be deemed "helpful" to the peace process. Anything for a quiet life for those politicians who are really interested in nothing other than the performance of their personal pension fund managers. Anything too taxing is off the agenda, which explains why today Northern Ireland is awash with multiple killers, some of whom have served as little as three months in prison.

Understanding terror is a complicated journey in itself. It's not just men in balaclavas lurking around dark alleys in search of unwary soldiers to use as targets of opportunity. We must come to understand the methods of terrorist tactics and training, think like them and sometimes even live like them. It may even take what I call "mind games" as I had to play with the six Loyalist bombers at Mount Norris in making them think I was P.I.R.A. Yes, it was dangerous; in fact some might say it was crazy, but it worked, and I have always said within the confines of the Irish "troubles" only the unexpected has any hope of working. Right now, at time of writing, the Royal Marines are preparing for yet another tour of duty in Northern Ireland and I know they are fully aware that professional paramilitarism is no longer just peopled by grubby little men in donkey jackets, looking to fire a shot and then run away. With the availability of any type of small arm you care to mention, para activity has become hugely sophisticated, if not indeed a learned way of life from a very early age and probably starting with a petrol-bomb production line from the garden shed. The learning curve travels an intimate training system; the weapons to use; the selection and effectiveness of bullet performance; the terrible wounds which can be inflicted by the careful selection of range, ballistics and bullet weight within the confines of an urban environment; the complicated deep black excursions into covert operations; the legal mine-fields of the dreaded rules of engagement, and so much more which impedes the security forces on a daily basis in their battle against terror.

For the security forces today, and for those very important younger people who wish to pursue a service career in the future, all the things we have looked at over Chapter 20 are of ongoing and paramount importance. Whether you fight urban paramilitarism in Belfast, Derry, The Congo, or Baghdad, this is the new kind of war/policing which I think the United States is going to find so very complicated, because waging war from B.52's at 40 thousand feet is a very long way from boots on the ground confrontation with terrorism, paramilitarism or guerrilla action, which is on an alarming roll today and growing fast. It took years of trial and error, combined with specialist operational training, to bring our own forces in Britain to the standard which is now emulated by the rest of the world, but the deep-rooted causes of terrorist activity are so complex and entrenched that if I had a choice, I would prefer to "yomp" the battlefield where a uniformed enemy could confront me in their thousands, rather than try to decipher the minds of the invisible madmen who visit Harrods or Canary Wharf with their hijacked vehicles and their deadly cargoes.

Food for thought for the Royal Marines in the dangerous years ahead.

CHAPTER 21

CORPORATE DECISION

When the British Task Force sailed for the Falkland Islands on 5th April 1982, the tactical details of Operation Corporate had yet to be finalised. The objective of the land forces was the capture of Port Stanley, and thus the most pressing decision to be taken was where exactly the men and equipment should be landed. At that point in time a Royal Marine Officer, Major Ewan Southby-Tailyour, who by good fortune had completed a yachtsman's survey of the islands three years previously, and has since explained to us how the force finally selected San Carlos, and how his Task Force Landing Craft Squadron got the men ashore. Weeks of painstaking research preceded the all-important decision to land the British Task Force at San Carlos in the Falkland Islands. But before we look at the hugely significant part he took in creating a plan towards finalising what became "Operation Corporate" let me first give you a brief history of this man with whom I served over many years within many differing elements of the Corps.

His father had been a Commandant General of the Royal Marines and before he was commissioned into the Corps in 1960, Ewan Southby-Tailyour had sailed 60,000 miles. During his career he has held commands in 45 Commando, with which he served on the Yemen border, 43 Commando and 42 Commando. He was lent to the French Commandos 'Joubert' and 'Clemenceau' in Toulon and, after attending their combat swimmer course, operated with the Foreign Legion. As the result of a series of actions in the Dhofar War with the Sultan's Northern Frontier Regiment he was awarded the Sultan of Oman's Bravery Medal for Gallantry.

His sea service included H.M.S. Wizard in the West Indies; command of the 1st Assault Squadron embarked in H.M.S. Anzio during the Radfan campaign, and command of the 4th Assault Squadron in

H.M.S. Fearless. He has served in the French submarine L'Estree, and the French commando carrier Arromanche, as well as H.M.S. Bulwark and H.M.S. Intrepid. From 1977 to 1979 he commanded Naval Party 8901 in the Falkland Islands. During the period of which we speak, he commanded 539 Assault Squadron, Royal Marines, based in Plymouth. He has completed 12 winters of training in the Arctic. The Falklands crisis found Major Southby-Tailyour in command of the Landing Craft Branch at Poole in Dorset. Seconded as Navigational Adviser to 3 Commando Brigade and the Commodore Amphibious Warfare, he was ordered to form the Task Force Landing Craft Squadron, for which he was made an O.B.E.

In his very first approaches to charting the Falklands, Ewan Southby-Tailyour (whom we will now refer to as E.S.T. during the remainder of our story for the sake of brevity), arrived on Port Stanley's public jetty in April 1978, having been transferred from H.M.S. Endurance in the MV Forrest. During that short journey of five minutes, he met the man of whom he had heard so much – Captain Jack Sollis, M.B.E., B.E.M. E.S.T. had been appointed to Naval Party 8901, as the Royal Marines Falklands Islands Detachment was known, and at last he was able to implement a plan he had nursed when first volunteering for the Royal Marines years before.

Since learning of his appointment in 1977, his researches showed that although the islands had been surveyed over the preceding 150-odd years by the Hydrographic Service, many of the charts were of little value to yachtsmen. He decided that he would interpret the Admiralty charts and record Jack Sollis's knowledge for any future visiting yachts. That the islands received very few visitors did not deter him for, at the worst, E.S.T. could produce handwritten notes for members of the Royal Cruising Club and Royal Yacht Squadron, to which he belonged. He also intended to study practical survival, the flora and fauna, and the history of the 300 or so known wrecks in the local waters.

There was, then, nothing ulterior in his motive for wanting to visit every landing craft beach or anchorage in the Archipelago, although it was easier to explain away his work by saying it was for military purposes. In fact, this was true for one set of beaches in the Campa Menta Bay area east of Salvador, where they placed a cache of arms, communications equipment and food for a stay-behind party in case of invasion. As he had also been trained as a landing-craft officer he was able to look at the landing places from complimentary angles; military landing beaches and yachtsmen's safe anchorages tend to possess the same characteristics.

During his year in the islands, three yachts called in. One in particular, a cement built gaffrigged ketch from Norway, was persuaded to allow him to put into practice many of the navigational disciplines necessary for sailing in this complicated collection of over 900 islands. It is difficult, even now, to calculate how many miles of coastline there are, but by the end E.S.T. estimated that he sailed 6,000 of them with the long-suffering Jack Sollis.

By the end of his tour, he had amassed about 1,000 slides and numerous watercolour sketches of the islands to supplement a notebook of over 100 pages of pencilled notes and sketches. It was this material that was so invaluable to him when the time came to advise the commanders on suitable landing places before San Carlos was finally chosen.

"We Have Lots of New Friends"

With these words, Port Stanley's telex operator broke the news of the invasion to a stunned and disbelieving Britain. It was 04.30 hours on 2nd April 1982 when 150 men of the Buzo Tactico – the Argentine Special Forces – landed by helicopter at Mullet Creek, a small inlet some three miles to the south-west of the Falklands capital Port Stanley. This was the beginning of the Argentine takeover of the Falkland Islands, and was followed up by the landing of over 1,000 more special troops and Marines. By 09.30 – five hours after the first Argentine soldier landed – the small, 80-man garrison of Royal Marines and others had surrendered. Just before noon the next day, 3rd April, the even smaller Royal Marine Force on South Georgia – some 800 miles east of Port Stanley – surrendered after one of the epic David and Goliath battles of our time. The shooting war had started.

The garrison of Royal Marines on the Falkland Islands was known as Naval Party 8901, and at 09.00 on 1st April (another grim coincidence), Major Mike Norman R.M. – in command of the

LDN: HELLO THERE: WHAT ARE ALL THESE ROUMOURS WE HEAR THIS IS LDN

FK: WE HAVE LOTS OF NEW FRIENDS
LDN: WHAT ABOUT INVASION RUMOURS

-2 APR. 1982

FK: THOSE ARE THE FRIENDS I WAS MEANING

LDN: THEY HAVE LANDED ?
FK: ABSOLUTELY

LDN: ARE YOU OPEN FOR TRAFFIC I.E. NORMAL TELEX SERVICE

FK: NO ORDERS ON THAT YET ONE MUST OBEY ORDERS

LDN: WHOSE ORDERS

FK: THE NEW GOVERNORS

LDN: ARGENTINA ?

FK: YES.

LDN: ARE THE ARGENTINIANS IN CONTROL.

FK: YES YOU CANT ARGUE WIT H THOUSANDS OF TROOPS PLUS ENORMOUS
NAVY, SUPPORT WHEN YOU ARE ONLY 18CO STRONG.
STAND BY PSE

TELEX FROM PORT STANLEY TO UK REPORTING INVASION

1982/3 Detachment – formally took over from Major Gary Noott R.M. and the 1981/2 troops job of defending the seat of Government. His job was to last for exactly 24 hours. At 15.30 that day, the two Majors were sent for by the Governor of the Falkland Islands, Mr. Rex Hunt, who showed them a signal from London. The message read, "An Argentine invasion fleet will be off Cape Pembroke at first light tomorrow. It is highly likely they will invade. You are to make the appropriate dispositions." The Government in London had been receiving reports from intelligence sources that Argentine Naval Forces conducted exercises at sea between 23rd and 28th March, which included a joint anti-submarine operation with the Uruguayan Navy in the River Plate estuary. Later reports showed that the fleet had sailed south from the main Argentine naval base at Puerto Belgrano on 28th March with Marines, soldiers and live ammunition on board. On 29th March it was known to be some 900 miles north of Port Stanley, and consisted of an aircraft carrier, four destroyers and an amphibious landing ship.

The Argentine Junta wanted to force Britain into conceding sovereignty over the islands in the South Atlantic. They needed something to restore their fast-deteriorating popularity at home, and violent demonstrations in Buenos Aires on the night of 30th March showed just how bad things were for them. By milking the diplomatic incident that had blown up over the landing of a party of scrap metal merchants at an old whaling station in South Georgia for all that it was worth, the Argentines had hoped to force concessions from the British Government at the Falkland sovereignty negotiations in New York.

While blowing up the scrap-metal merchant incident, the Junta also strengthened their hand militarily by despatching warships to South Georgian waters and later by putting their task force to sea, equipped to invade the Falkland Islands themselves.

The Junta had been badly briefed. They thought that Britain would back down, rather than risk a war. They also thought that the British would give in, when faced with superior numbers.

It seems that the actual orders to invade were sent to the Task Force Commander quite late on – only one or two days before 2nd April. The ease with which they were carried out shows that the Argentines must have rehearsed the action in advance, and had worked out very detailed plans. Admiral Carlos Busser was there to command the 4,500 invasion troops.

Major Mike Norman had the thankless task of defending the islands against this overwhelming force. To help him, he had just 43 men from the 'new' Naval Party 8901, 25 from the 'old' party, and 12 sailors from the ice patrol ship H.M.S. Endurance. Nine men from the 'old' party had been sent with Lieutenant Keith Mills to South Georgia, there to impose the will of Her Majesty's Government on the scrap dealers. Norman had an impossible job, and he knew it. The task he set himself, therefore, was to hold out for as long as possible. Norman reasoned that the enemy would go for Port Stanley and neutralise the defences there so that forces could be landed on the airstrip and in the harbour. Accordingly, he organised his defences around the airfield and to the east of Stanley itself. His first priority was to make sure the runway could not be used; he ordered vehicles to be parked on it, with a single small section of men (5) covering the obstacles from the south. An Observation Post was set up to the east of Yorke Point, and a machine gun group was to be positioned overlooking part of Yorke Bay – one possible landing site – which was also obstructed by coils of barbed wire. The machine gunners were given two motorcycles for a quick getaway, and a canoe was also hidden as part of an emergency escape plan.

Just south of No. (5) section's position, the road from the airfield to Port Stanley makes a right-angled turn at Hooker's Point. It was here that No. (1) section was placed. To the west of them, at the old airstrip, was No. (2) section. Their firepower was to be increased by the addition of an 84 mm Carl Gustav anti-tank weapon and its two-man team. They were also to have light 66 mm anti-tank missiles. Farther west still were the VOR directional beacons for the airfield's approaches. No. (3) section was positioned nearby; their task would be to delay the enemy for as long as possible before withdrawing.

Across Port Stanley Bay, at Navy Point, was No. (4) section's location. They were to have the other 84 mm Carl Gustav MAW, and were given the thankless task of engaging any enemy landing craft or shipping which tried to get through the narrows. They had a Gemini inflatable boat for a quick return to Government House if they were needed there.

To the south of Port Stanley, on Murray Heights, No. (6) section was placed to give early warning of any Argentines approaching from the south of the town. Another O.P. was to be located to the west of No. (6) section's position on Sapper Hill and the sole Marine manning it, Mike Berry, was to be equipped with a third motorcycle. The main headquarters was to be at Government House with Major Noott acting as the Governor's adviser. Major Norman was to be at Look Out Rocks, on the edge of the town to the south-east, commanding the troops on the ground. The motor vessel MV Forrest was sent to sea to keep a radar watch in the waters off Port William to the north. Unfortunately, the party's 81 mm mortar was damaged and so couldn't be used, but an extra machine gun was given to No. (2) section at Hooker's Point.

During the night, the garrison was joined by two islanders, Jim Fairfield – a former Royal Marines Corporal – and Bill Curtiss. The Marines' barracks at Moody Brook Camp were vacated by 02.00 on the morning of 2nd April (the same day as Ewan Southby-Tailyour was summoned to Headquarters Commando Forces by Brigadier Julian Thompson), and all defensive positions were occupied. The light was switched off at Pembroke Point lighthouse, and Bill Curtiss put the VOR beacon out of action. The tiny British force was as ready as it could be to receive the Argentine invaders. All that remained was for the 37 Argentines on the island to be rounded up and placed under guard.

02.30. "Contact!" It was the Forrest – the enemy were on their way. An hour later, at 03.30, it was clear that a large fleet was manoeuvring off Cape Pembroke. There was not a man in the defending force who did not feel the strain of waiting, but they knew they wouldn't have to wait for long. Norman had concluded his final briefing with these words: "Remember, you're not fighting for the Falklands. This time you're fighting for yourselves."

04.30. "Contact!" It was the O.P. on Sapper Hill. Helicopters near Mullet Creek in the general direction of Port Harriet. What were they doing there? The answer came at 06.15 when firing and explosions were heard from the direction of Moody Brook. It was the Buzo Tactico, shooting up the Marines' barrack block there. The Argentines went in hard with sub-machine guns and fragmentation and phosphorus grenades in an attempt to catch the Marines in bed

– which made a mockery of subsequent claims that the Argentines had been trying to spare lives. What immediately appalled Major Norman, however, was that he had been "wrong footed". They were coming from the west, and all his forces were spread out to the east. Numbers (1) and (5) sections were immediately ordered to make for Government House, and Major Norman returned there himself post haste. The first attack against the headquarters came in at 06.15 and the Argentines managed to get very close to the house – indeed three of them got into the maids' quarters where they hid in the loft – but the attack was beaten off by rifle and machine gun fire.

About a quarter of an hour later – at 06.30 – the O.P. at Yorke Point and No. (2) section at the airstrip reported that craft were landing at Yorke Bay. Some minutes later they radioed that about 18 armoured personnel carriers (APC's) were ashore and advancing towards Port Stanley. These were American-made LVTP-7 "Amtracks", equipped with .30 in. calibre machine guns. As they advanced down the airport road, they were engaged by No. (2) section under Lieutenant Bill Trollope R.M. The leading vehicle was stopped by a 66 mm. LAW (light anti-armour weapon) missile fired by the armourer, Marine Gibbs, which hit the passenger compartment. An 84 mm MAW round fired by Marines Brown and Best exploded against the front of the Amtrack. No-one got out. The other vehicles deployed their troops and opened fire with their machine guns.

Beating a prudent retreat, the Marines made their way back to Government House where a fierce battle had been raging for some hours as the Buzo Tactico tried to storm it. During the heat of the fighting, a radio message was received from No. (4) section near Navy Point saying that it had three targets to engage and what were the priorities. "What are the targets?" asked Norman. "Target No. 1 is an aircraft carrier; target No. 2 is a cruiser; target ......" The radio went dead. The section made good its escape in the Gemini inflatable, however, and remained undetected for four days after surrender.

The battle had, by this stage, turned into a sniping contest. It was now daybreak and more than one Marine reckoned that they could hold out against the 600 or so men surrounding Government House. But as Governor Rex Hunt telephoned around Port Stanley to check with the inhabitants on the progress of the Argentine advance, the unwelcome news arrived that the Amtracks were on their way. They could sit well out of range of the Marines' fire and, if they wanted, raze Government House to the ground. Norman knew that there was no longer any chance of a breakout, so he suggested that Hunt talk to the Argentines. The Governor agreed reluctantly, but did not intend to surrender to the Argentines. He had no choice, however. Vice Commodore Hector Gilobert, an Argentine who ran L.A.D.E., the civil airline supplying the Falklands, was contacted at his home in Stanley (there hadn't been enough time to arrest him), and he agreed to act as go-between. At this moment, the three Argentine commandos who had been hiding in the maid's bedroom decided it was time to make a move. Crawling cautiously from under the bed, they moved towards the door. In the room directly underneath was Gary Noott. Without a moment's hesitation he fired several rounds from his S.L.R. rifle through the ceiling followed, when he heard the men crying out in shocked surprise, by several more shots. Ten seconds later the Argentines tumbled down the stairs, unhurt, with their arms in the air, to be taken prisoner.

When Hunt met with Admiral Busser at Government House, he told the Argentine that he was unwelcome, and invited him to leave. The Admiral politely declined, pointing out that he had some 2800 men ashore and 2000 more on ships. The logic was inescapable. At 09.25 Hunt ordered the Marines to lay down their arms.

The Falklands Options

As the Task Force sailed south and the troops prepared themselves for the battle to recapture the Falklands, senior military planners, drawn from every branch of Britain's armed forces, gathered together in a series of exhausting meetings to thrash out a war-winning strategy. Their chief concern was to settle upon a suitable location for the initial amphibious assault – potentially the most dangerous part of the proposed operation. A number of factors had to be considered: the geography of the islands and the differing needs of the land, sea and air forces. Official intelligence on the Falklands was extremely scanty and much of what was available was out of date. However, a Royal Marines officer, Major Ewen Southby-Tailyour, was able to provide the missing information. A keen sailor, he had been a frequent visitor to the islands before the war and had spent much of his spare time plotting the Falklands' extensive coastline. Armed with Southby-Tailyour's detailed charts and diaries, the Task Force's planning committee was able to look at a number of possible landing sites. One by one, each option was either rejected or put aside for further consideration. Finally, after many hours of debate, a compromise site was agreed: the troops would be put ashore at San Carlos.

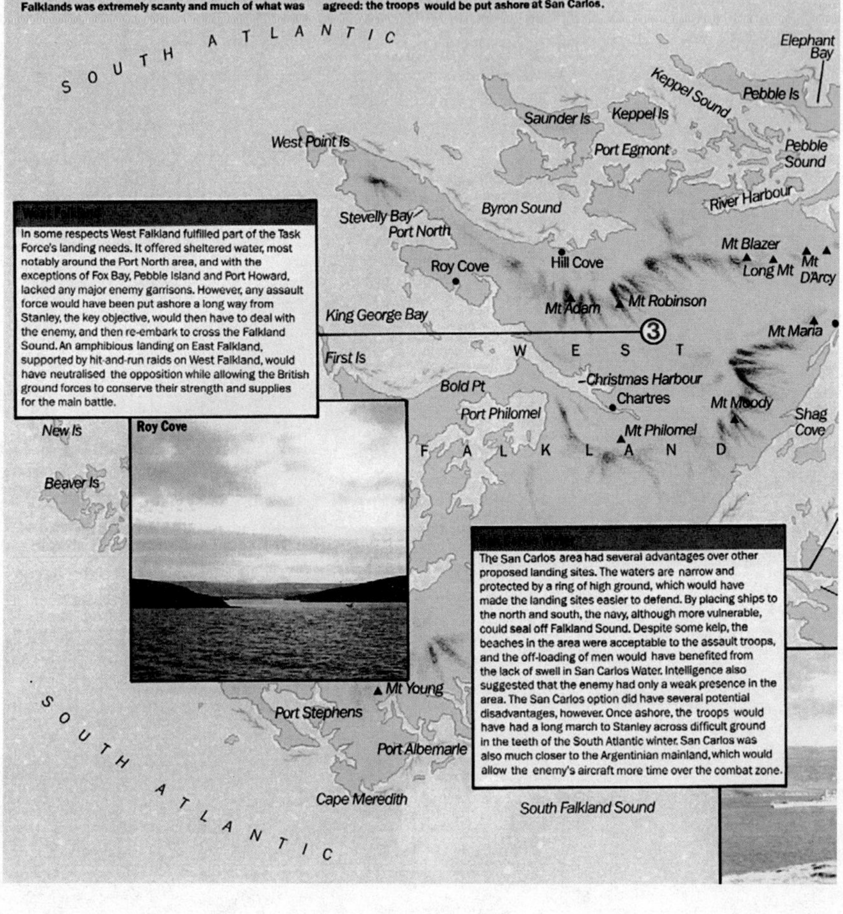

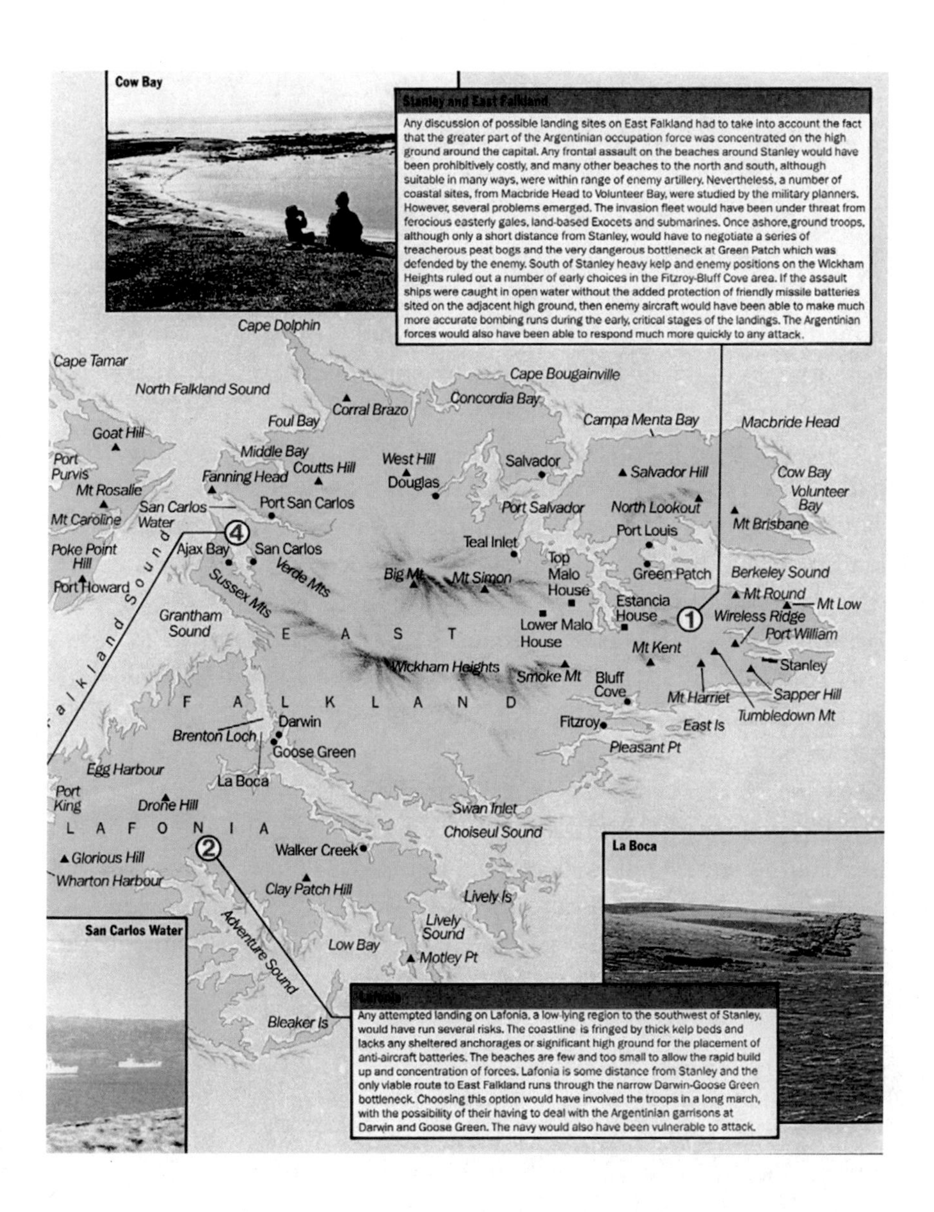
Cow Bay
Stanley and East Falkland
Any discussion of possible landing sites on East Falkland had to take into account the fact that the greater part of the Argentinian occupation force was concentrated on the high ground around the capital. Any frontal assault on the beaches around Stanley would have been prohibitively costly, and many other beaches to the north and south, although suitable in many ways, were within range of enemy artillery. Nevertheless, a number of coastal sites, from Macbride Head to Volunteer Bay, were studied by the military planners. However, several problems emerged. The invasion fleet would have been under threat from ferocious easterly gales, land-based Exocets and submarines. Once ashore,ground troops, although only a short distance from Stanley, would have to negotiate a series of treacherous peat bogs and the very dangerous bottleneck at Green Patch which was defended by the enemy. South of Stanley heavy kelp and enemy positions on the Wickham Heights ruled out a number of early choices in the Fitzroy-Bluff Cove area. If the assault ships were caught in open water without the added protection of friendly missile batteries sited on the adjacent high ground, then enemy aircraft would have been able to make much more accurate bombing runs during the early, critical stages of the landings. The Argentinian forces would also have been able to respond much more quickly to any attack.
Cape Dolphin
Cape Tamar
North Falkland Sound
Cape Bougainville
Concordia Bay
Corral Brazo
Foul Bay
Campa Menta Bay
Macbride Head
Goat Hill
Middle Bay
West Hill
Salvador
Port Purvis
Fanning Head
Coutts Hill
Douglas
Salvador Hill
Cow Bay
Mt Rosalie
Volunteer Bay
San Carlos Water
Port San Carlos
Port Salvador
North Lookout
Mt Caroline
Mt Brisbane
Port Louis
Poke Point Hill
Ajax Bay
San Carlos
Teal Inlet
Top Malo House
Verde Mts
Big Mt
Mt Simon
Green Patch
Berkeley Sound
Port Howard
Sussex Mts
Mt Round
Estancia House
Mt Low
Grantham Sound
Lower Malo House
Wireless Ridge
Port William
E A S T
Mt Kent
Falkland Sound
Wickham Heights
Smoke Mt
Stanley
Bluff Cove
Sapper Hill
Mt Harriet
F A L K L A N D
Tumbledown Mt
Darwin
Fitzroy
East Is
Brenton Loch
Goose Green
Pleasant Pt
Egg Harbour
La Boca
Port King
Drone Hill
Swan Inlet
L A F O N I A
Choiseul Sound
Glorious Hill
Walker Creek
Wharton Harbour
Clay Patch Hill
Lively Is
San Carlos Water
Lively Sound
Adventure Sound
Low Bay
Motley Pt
La Boca
Bleaker Is
Any attempted landing on Lafonia, a low-lying region to the southwest of Stanley, would have run several risks. The coastline is fringed by thick kelp beds and lacks any sheltered anchorages or significant high ground for the placement of anti-aircraft batteries. The beaches are few and too small to allow the rapid build up and concentration of forces. Lafonia is some distance from Stanley and the only viable route to East Falkland runs through the narrow Darwin-Goose Green bottleneck. Choosing this option would have involved the troops in a long march, with the possibility of their having to deal with the Argentinian garrisons at Darwin and Goose Green. The navy would also have been vulnerable to attack.

CHAPTER 22

UNFOLDING THE BATTLE PLAN

Much later as the Task Force sailed south and the troops prepared themselves for the battle to recapture the Falklands, senior military planners, drawn from every branch of Britain's armed forces, gathered together in a series of exhausting meetings to thrash out a war plan. Their chief concern was to settle upon a suitable location for the initial amphibious assault – potentially the most dangerous part of the proposed operation. A number of factors had to be considered; the geography of the islands and the differing needs of the land, sea, and air forces. Official intelligence on the Falklands was extremely scant and much of what was available was out of date. However, a Royal Marines officer, Major Ewan Southby-Tailyour, was able to provide the missing but vital information. A keen sailor, he had been a frequent visitor to the islands before the war and had spent much of his spare time plotting the Falklands' extensive coastline. Armed with E.S.T.'s detailed charts and diaries, the Task Force's planning committee was able to look at a number of possible landing sites. One by one, each option was either rejected or put aside for further consideration. Finally, after many hours of debate, a compromise site was agreed; the troops would be put ashore at San Carlos.

"San Carlos Water"

The San Carlos area had several advantages over other proposed landing sites, which we will look at in turn. The waters are narrow and protected by a ring of high ground, which would have made the landing sites easier to defend. By placing ships to the north and south, the Navy, although more vulnerable, could seal off Falkland Sound. Despite some kelp (large brown seaweed), the beaches in the area were acceptable to the assault troops, and the off-loading

Water. Intelligence also suggested that the enemy had only a weak presence in the area. The San Carlos option did have several potential disadvantages, however. Once ashore, the troops would have had a long march to Stanley across difficult ground in the teeth of the South Atlantic winter. San Carlos was also much closer to the Argentine mainland, which would allow the enemy's aircraft more time over the combat zone.

"West Falkland"

In some respects West Falkland fulfilled part of the Task Force's landing needs. It offered sheltered water, most notably around the Port North area, and with the exceptions of Fox Bay, Pebble Island and Port Howard, lacked any major enemy garrisons. However, any assault force would have been put ashore a long way from Stanley, the key objective, would then have to deal with the enemy and then re-embark to cross the Falkland Sound. An amphibious landing on East Falkland, supported by hit-and-run raids on West Falkland, would have neutralised the opposition while allowing the British ground forces to conserve their strength and supplies for the main battle.

"Stanley and East Falkland"

Any discussion of possible landing sites on East Falkland had to take into account the fact that the greater part of the Argentinian occupation force was concentrated on the high ground around the capital. Any frontal assault on the beaches around Stanley would have been prohibitively costly, and many other beaches to the north and south, although suitable in many ways, were within range of enemy artillery. Nevertheless, a number of coastal sites, from Macbride Head to Volunteer Bay, were studied by the military planners. However, several problems emerged. The invasion fleet would have been under threat from ferocious easterly gales, land-base Exocet missiles, and submarines. Once ashore, ground troops, although only a short distance from Stanley, would have to negotiate a series of treacherous peat bogs and the very dangerous bottleneck at Green Patch, which was defended by the enemy. South of Stanley, heavy kelp and enemy positions on the Wickham Heights ruled out a number of early choices in the Fitzroy - Bluff Cove area. If the assault ships were caught in the open water without the added protection of friendly missile batteries sited on the adjacent high ground, then enemy aircraft would have been able to make much more accurate bombing runs during the early, critical stages of the landings. The Argentinian forces would also have been able to respond much more quickly to any attack.

"Lafonia"

Any attempted landing on Lafonia, a low-lying region to the south-west of Stanley, would have run several risks. The coastline is fringed by thick kelp beds and lacks any sheltered anchorages or significant high ground for the placement of anti-aircraft batteries. The beaches are few and too small to allow the rapid build up and concentration of forces. Lafonia is some distance from Stanley and the only viable route to East Falkland runs through the narrow Darwin – Goose Green bottleneck. Choosing this option would have involved the troops in a long march, with the possibility of their having to deal with the Argentinian garrisons at Darwin and Goose Green. The Navy would also have been vulnerable to attack.

Major Ewan Southby-Tailyour reported to Brigadier Julian Thompson in Headquarters Commando Forces at 20.00 hours on 2nd April 1982 and was asked to tell all he knew about the Falkland Islands. It took E.S.T. a moment to realise that if he parted with all his expertise and long-studied geographical knowledge essential to any logistical assault on a place such as this, there was then a chance that he just might be left behind in England when the Task Force sailed. Well, the warrior within him would have none of this, but he need not have worried because, thankfully, the General agreed to his request for an appointment to the command structure for the duration of whatever lay ahead. That night E.S.T. began the first of 69 talks on the subject, using a selection of his 1,000 slides, numerous sketches and hand-drawn charts. The first presentation covered various aspects of life and conditions in the Falklands and ranged from the politics and economics to the shores, the wrecks, the weather, survival, the ecology and the amazing wildlife.

For that initial briefing he produced a small map of the islands on which he had scribbled comments alongside 16 of the most likely areas for amphibious operations. At this point both he and the General had no idea of their tasks, but whatever else was decided in London they knew that they would have to plan a landing with Stanley (the capital) as their final objective. It was assumed, then, that the Brigade would have the use of an aircraft carrier of sorts to provide the required helicopter spots from which to launch the initial assault.

After the Brigade Headquarters embarked on H.M.S. Fearless on 6th April, E.S.T. began the task of collecting and collating information and photographs from every Marine who had served in Naval Party 8901 in the islands and who could help to build a winning picture. From these memories and from the answers to hundreds of questions signalled back to Falkland Islanders living in England, the team built up a picture of the area over which they were to operate. But the picture that they painted of the islands, it had to be said, was not very encouraging from the military point of view. They explained that as a place to live, the islands were not unpleasant, with a climate best described as sunny, damp, cool, and very windy. As an active service environment, the single most important problem was the rate of change of the weather, which could be a nightmare for the military. They explained that the "going" for ground troops could be very difficult, with 85 per cent of the 4,700 square miles covered by peat bog and with no natural cover from hedges or trees. Whether or not the routes led through the mountains or across the plains, the approaches to Stanley would be arduous. The military needed as short an approach to the objective as possible, and yet a frontal assault on the local beaches was out of the question. The British do not do things that way. The Argentinians were influenced by American thinking and, believing that we were equally influenced, they expected us to land on the east of East Falkland. In this they miscalculated the military thinking, the subtlety, fitness and endurance of the incoming Task Force.

Landing near Stanley would have made the Force vulnerable to the easterly gales, and to Exocet and submarine attack. It also would have guaranteed heavy civilian casualties, a further unacceptable factor in the planning. Luckily, the enemy had placed his defences in accordance with U.S. military philosophy and even after our landings had taken place he was reluctant to re-deploy in strength, believing that the landings were a diversion.

As the team understanding of the defensive positions became clearer, they began looking at beaches and options further away from Stanley and the enemy heavy artillery. Looking north of Stanley they studied the Volunteer and Cow Bay areas, and indeed, the whole of the Macbride Head peninsula. However, the land approaches to Stanley were not good and there was a bottle-neck between Salvador Waters and Berkeley Sound which was certain to be heavily reinforced. If they had landed in this area they would have had to fight a fixed battle at Green Patch, an area which would have taken the place of Goose Green in British military history. But there, unlike at Goose Green, they would have been forced into battle.

To the south of Stanley there was little that presented itself as sensible from either the military or naval points of view. The beaches were all small and far apart and the approaches were difficult, with numerous kelp reefs. The enemy would still have a clear run at the amphibious ships, the ground was low and not good for the setting up of Rapier anti-aircraft defences, and the enemy would have been able to observe all from the Wickham Heights.

The whole of Lafonia was discounted by the amphibious forces for the same reasons, with the added factor of a terrible "yomp" with full kit across low country. Again there was the disadvantage of a bottle-neck – Goose Green. Within the team, the Brigadier was conscious that the march towards Stanley would be arduous enough without any unnecessary battles on the way. At one stage they were tasked with looking at Stevelly Bay in Port North For, a few miles behind the beachhead, there was an ideal area to build an airstrip from which Harriers, and eventually even Hercules transports could operate. It would have taken a long time to construct and was significantly closer to the Argentinian air bases. It was not an option that met with the approval of the amphibious forces. They had to bear constantly in mind that Stanley was their goal and that any extra battle or option would deplete their limited stocks of logistics and endurance, thus making the fight for the capital of Stanley more difficult.

The San Carlos option was formed slowly as their attentions drifted away from Stanley. Although it had been discussed form the earliest days, so had many other areas. E.S.T. first entered it in his diary as a specific area deserving their closest scrutiny on 17th April.

"Arrived Ascension Island. Briefed the Commander-in-Chief Fleet on board H.M.S. Hermes. Admiral Fieldhouse gave us the aims.....(the fourth one was: 'Plan to land on the Falkland Islands with a view to re-establishing British administration'. Returned on board H.M.S. Fearless. Brain-stormed landings at San Carlos with Brigadier."

San Carlos had both advantages and disadvantages for the military force and the amphibious ships. The Royal Navy liked the location because it would be clear from Exocet attacks, and the submarine threat could be contained by blocking each end of Falkland Sound. (Although there remained a worry that one of the Argentinians' very quiet German-built submarines might have positioned itself on the bottom, waiting for just such a British action). The anchorages were sheltered from all quarters of wind and swell, which meant that the offload could take place uninterrupted by nature. The ground forces could set up a circle of Rapier to defend the ships and beachhead. But for the Royal Navy the choice of San Carlos would also mean a much longer transit of the Total Exclusion Zone, much of which would have to be undertaken in daylight. There were also worries of mines in the entrances to Falkland Sound and San Carlos Water – although this problem was constantly ignored! Again, while enemy aircraft (having evaded the Harriers and Rapier on their approaches to the anchorage) would have less than three seconds to select and engage a target after flying over the surrounding hills, the ships would have as little time to operate their defensive systems. This would compel the Navy to have ships out in the sound and to the north and west of the islands to give early warning.

The Royal Marines liked San Carlos as there were four acceptable beaches for landing craft, each one lying beneath a high feature. This would ensure good all-round defence of the anchorage on every side. The complicated offload and the setting up of the initial supply depots could be undertaken with some security from ground attacks. San Carlos was, therefore, ideal in this respect. There were, however, some distinct military disadvantages that would become more relevant after the beachhead had been secured. The port was as far from Stanley as was possible on East Falkland, (except for Low Bay which, for some reason not even now understood, always seemed to be the preferred option of the Battle Group Staff); and this would mean a long approach march during which the enemy could re-deploy his defences. San Carlos, lying to the west, was also closer to the Argentinian mainland and enemy aircraft, although near the limit of their endurance, would be able to attack targets of opportunity. The port was also some distance from the ships of the Battle Group and their embarked Harriers to the east of Stanley, and the aircraft would have a very limited time on target during any defence action.

From the point of view of E.S.T., San Carlos was a good place, for although the beaches were not excellent he did know them all and could advise accurately on their characteristics. He could take landing craft through the kelp reefs and to his reckoning the Argentinians, if they had thought that the British Forces would land in the area, would not have bothered to mine the kelp, assuming that the kelp banks would be a natural barrier to incoming troops and equipment.

Although San Carlos was chosen as representing the place with the least disadvantages, the final decision was not made until very late. On 2nd May, for example, the Cow Bay/Volunteer Bay area seemed the most likely place. On 4th May E.S.T. typed out a complete navigational guide to Salvador Waters, and also on that day he was asked to brainstorm the coast east of Cape Dolphin. This was not difficult from his point of view, as it was probably the worst stretch of coastline in the archipelago for a landing.

On 6th May he was invited to re-study the beaches along the coast east from Salvador. Although open to the unpredictable northerly winds and not liked by the Royal Navy, there were distinct military advantages as the series of beaches would ensure the simultaneous landing of three battalion-strength units. As a result, an S.B.S. team recced the beaches and confirmed his previous observations. This was very good for the credibility of E.S.T., for until then no-one had ascertained the standard, or reliability, of his surveys. On 8th May he briefed the S.A.S. for their approach to Pebble Island. They were to set up a terminal guidance party in advance of a helicopter-borne S.A.S. raid designed to destroy Argentinian aircraft on the island's airstrip.

His diary for 8th May also revealed that they were: "Now looking seriously at the Port San Carlos area." The next day, in addition to discussing the San Carlos option, the Brigadier and he again mulled over Salvador Waters. They also heard that the S.B.S. recce of San Carlos had found minimal enemy activity, a fact which helped them in making up their minds. Of course, any decision had to be taken in London at the highest level, but E.S.T. hoped that they would listen to their advice. His diary for 10th May read: "After dinner, worked on the Navigational Plan for the Commodore of Falkland Sound and briefed the C.O. S.A.S. on the navigational problems in Brenton Loch and La Bocha."

In the middle of all this planning he tried to keep a balance, noting that the first Wandering Albatross had been sighted at 30∞50'S, 26∞00'W, and the first Wilson's Petrel at 31∞56'S 26∞00'W. He was also sent off Fearless's bridge for encouraging the Officer of the Watch to keep him informed of bird sightings, instead of submarines and aircraft. However, E.S.T. painted a silhouette chart of all the Southern Ocean whales and dolphins and pinned this alongside the ship and aircraft recognition posters. He felt particularly strongly that those who had never visited this part of the world should appreciate all the reasons for securing its freedom.

The Brigade Commander gave his orders for the landings at San Carlos in the wardroom of Fearless on 13th May. E.S.T. opened the proceedings with the "ground" paragraph and retraced everything he had told all of the land forces during the previous weeks of individual lectures. He then homed in on the Amphibious Operations area. Later he flew around the Task Force military units and gave even more detailed briefs.

The codeword ordering the landings to take place in San Carlos Water was received by the Brigadier at 11.25 hours (local) on 20th May as he watched a couple of his team play a game of backgammon. They were 130 miles east of Stanley. Many books have been written on the main landings and the events right through to the Argentinian surrender, so I will not go into monumental detail as "Of the Hunter Class" is meant to be a recording of historic events or, if you like, "Reflections of Valour", which warranted being brought to the fore and recorded briefly for the benefit of all who want a finer understanding of Marine-Naval co-operation. As the story of the main landings is well documented, we will simply outline events as they unfolded. 2 Para and 40 Commando were taken by the assault squadron from the initial anchorage in Falkland Sound around

Chancho Point. It was dark (22.30 local time on 20th May), and the run should have taken one hour from the line of departure (the nautical equivalent of the Start Line), which stretched between Fanning Head to the north and Chancho Point to the south. E.S.T. had planned the approach at six knots with all 16 craft in the first wave. This gave him time in hand to make the first beaching with half of the boats at 23.30. Inevitably, there were problems, and they left the L.O.D. an hour late. This was more than they could make up by using their top speed of nine and a half knots, but it was vital that they caught up this time as 2 Para, in particular, had a very long march south to reach their objective on Sussex Mountain by dawn. They steamed at full speed and resisted the temptation to cut corners and head straight down San Carlos Water. Instead, E.S.T. kept to his plan of hugging the western shore; by doing so they kept to the edge of the kelp banks in shallow water, which was a route he was sure no enemy would have expected them to take. Immediately opposite 2 Para's beach (in Bonners Bay) they turned to port and made the final dash across the open estuary, putting the battalion ashore exactly 45 minutes late. Ten minutes later, 40 Commando, having waited in Ajax Bay, was landed to the north of the settlement at Pony's Valley beach. The two halves of the landing craft squadron then rejoined off Ajax Bay and returned the way they had come before splitting again to collect 3 Para and 45 Commando from their ships. The Royal Marines were then taken back down the same route and landed at Ajax Bay, while E.S.T. took the para battalion across the line of departure, past Fanning Island and through the kelp beds to Rabbit Island, for a wet landing on Sand Beach as dawn broke. Throughout, there had been no opposition, although they had expected to be opposed by an enemy company. As a landing-craft and navigational operation it was carried out as faultlessly as if it had been for a General's inspection; which, in the words of E.S.T. is I suppose, why we have them! On that, I'm not too sure if he meant "Generals" or "Inspections"!

Subsequently, the landing and raiding craft were tasked by the Commodore's staff to assist in the offloading of the logistic and passenger ships and in the setting up of two major beach units; one at Bonners Bay beach for the stocking of supplies required forward, and the other at Ajax Bay for the establishing of the Brigade Maintenance Area and the Commando Logistic Regiment.

During these early days the craft were tasked for such work as towing the crippled H.M.S. Argonaut into the comparative safety of the estuary. The frigate was unable to move or fight due to two bombs that, although they had not exploded, had caused considerable damage to her weapon and propulsion systems. Moving Argonaut was a fascinating manoeuvre carried out under air attack and in very strong winds. They secured one L.C.U. to each side amidships to provide power, with a third L.C.U. steering from ahead on a long tow. The L.C.U.'s were also instrumental in rescuing most of the crew from H.M.S. Antelope as she burned, the coxswains having to be ordered to leave her side for their own safety. Later they were to provide similar gallant assistance to the L.S.L.'s Sir Galahad and Sir Tristram. All the raiding craft were employed in the insertion and removal of S.A.S. and S.B.S. patrols. These tasks with their highly valuable "cargos" placed great responsibility on young Marines who were working singly or in pairs, commanded only by a corporal. One of the trickiest navigational tasks was the re-insertion of 42 Commando from the Cerra Montevideo Hills to the east of Port San Carlos. They did not want this move to be known by the enemy and so the only way was to lift them out by night, using the San Carlos River than runs in towards the settlement from the east. The north bank of this river was held by 3 Para and the south by the enemy; at least, it had not yet been cleared by us. The round trip was six miles along a river that none of them knew and that E.S.T. had surveyed only briefly in 1979. They reckoned they could get away with it by night, but the imagination ran wild when considering the consequences of having a fully laden craft stranded at dawn and at the mercy of the enemy's Pucarãs.

The operation took all night, during which time they hit the bottom three times. Each newly discovered shoal was carefully plotted for the next trip. E.S.T. noted in his diary that much of the navigation was "conducted by instinct".

On 24th May, Colonel 'H' Jones of 2 Para discussed with E.S.T. the deployment of his battalion down Brenton Loch for a full-scale opposed assault landing right onto the Argentinian flank. Sadly, as it turned out, after three hours of careful planning, he decided that he could not risk the low odds that E.S.T. was offering for navigating the Loch by night. It was thought likely that at least one craft would be stranded by dawn on any one of the numerous shoals. Although E.S.T. had navigated the Loch often enough in peacetime it had always been in daylight, using a radar for accurate position fixing. In some places the deep water track was so narrow and erratic that losses would have been almost inevitable.

'H' decided, in his own words to "tab it". How very different things might have been if he had used the landing craft as offered by E.S.T.

During 3 Commando Brigade's move towards Stanley along the northern flank, a number of the L.C.V.P.'s were used in the role of minesweepers in the entrance to Salvadore Waters. This was one of the more hazardous undertakings and was, as usual, conducted by very young and junior N.C.O.'s who more than earned their "Mentions in Despatches".

It was probably the deployment of the Scots Guards to Bluff Clove on the night of 5th/6th June that caused the most worry to E.S.T. and the Task Force Landing Craft Squadron. Because of the lack of helicopters (and, he was very truthful in saying, the fact that the troops were not as fit as those who had yomped along the northern flank), it was decided to lift the Scots and Welsh Guards forward by landing craft. The land-based Exocet threat prevented the L.P.D.'s from sailing further east than Lively Island, and so on a beautiful calm night four of H.M.S. Intrepid's L.C.U.'s were launched into Lively Sound. E.S.T. had particularly asked to be escorted further east by the accompanying frigate but this, and Harrier Combat Air Patrol (C.A.P.) cover for dawn, was denied him. He felt that the possible loss of a whole battalion at that stage of the campaign outweighed the risk of a frigate coming under attack. However, it was difficult to persuade the L.P.D. Captain of the navigational and military dangers of a seven-hour passage in unprotected and unarmed open boats, equipped with the crudest of navigational aids and off a coastline that might have been held by the enemy. His three-year-old memory of the coast was just about all E.S.T. had, as his own charts of Lively Island and Bluff Cove had been left behind. He had joined the ship expecting to take the men into the entrance to Brenton Loch to save them a few miles walking towards Goose Green. He left the Captain on the bridge in little doubt about the chances of success. (On subsequent journeys, Fearless took them to the south of Elephant Island, which made their task considerably easier).

Sadly, his fears did not prove groundless. The weather deteriorated rapidly; the radars (on the occasions that E.S.T. felt it safe to use them) were inadequate; they were shelled or mortared from ashore, and were star-shelled by H.M.S. Cardiff just when they thought that they had passed undetected. She had not been informed of their presence in the area and, indeed, E.S.T. had been told that there were no British ships on the "gunline" that night.

Do I detect the merest touch of "Murphy's Law" here? However, E.S.T. had presumed that this was the reason for not giving him any identification codes or frequencies. Seeing her, and another vessel, approaching them at speed in the dark he assumed that they were the Argentinian gunboats of which he had been warned, so he decided that, as he knew the waters better than they, he would take his small flotilla into the comparative safety of shoal waters. In the rising wind they were able to make only three knots and as it was clearly a futile gesture, they hove to when requested. The lead ship then flashed the one word "Friend", and E.S.T. ordered his signaller to reply, "To which side!" At that the ships disappeared over the horizon, failing to appreciate the flotilla's need for an escort.

Digression: Over the years and reading of many such incidents on the lighter side of military ops., I often wondered what the late great David Niven would have thought and how he would have put all of them together as a collection of stories in his very specialised but inimitable way as a former officer with a sense of humour as free as the wind. With the material that's available today, the way is wide open for a successor to Niven's "The Moon's a Balloon." Perhaps E.S.T. Well, there must be someone out there!

To get back to the problem at hand, on closing the coast south of East Island E.S.T. had hoped to be able to transit East Road, but the pass was a mass of turbulent water. As dawn was imminent he reluctantly decided to head even further east to round the extensive kelp banks lying only three miles or so from Sapper Hill. They found the 60 ft wide gap at Bluff Cove just before daybreak and entered against 70 knots of wind to land the Scots Guards exhausted but safe. They had endured seven unspeakable hours huddled under their ponchos on the cold and wet open well decks, knowing that they were prey to the weather and the enemy and that there was nothing they could do about either. E.S.T. marvelled at them and their irrepressible humour as they struggled off the craft towards their defensive positions in the equally sodden peat. The squadron continued to support the two brigades on both flanks, with the raiding craft conducting particularly hazardous operations to the south-east of Salvador Waters and on the islands close to Stanley. It was they that carried the S.A.S. on the abortive attack on Wireless Ridge, when all craft were burnt out as a result of enemy fire.

All in all, it was a "good" war for the Task Force Landing Craft Squadron. When coupled with its work in north Norway it would have been difficult for the Admiralty to refuse the Royal Marines' request to form the squadron formally. In his "Report of Proceedings" after the conflict, E.S.T. observed that they had learnt few new lessons because the operating conditions, loads carried, geographical and climatic problems had been similar to those in Norway. Their experiences in the Falklands had underlined the fact that their Arctic training was fully relevant to worldwide operations.

CHAPTER 23

MAKING COMPARISONS

Brigadier Thompson had a positive strategic plan worked out for mounting a major assault across the Falklands from San Carlos to Port Stanley. But his plan had not taken into consideration the Exocet missiles winging his way. Nor had he reckoned with career-fretful politicians back in the U.K. Some say this war was a godsend for Margaret Thatcher, whose popularity was flagging for all sorts of reasons, and two days before Argentine forces occupied the Falklands, ten thousand irate citizens took to the streets of Buenos Aires to vent their anger against the highly unpopular Junta and its leader, General Leopoldo Galtieri who also found himself in desperate need of a huge distraction from his failed policies. I have often asked myself if this was another case of "Events, dear boy, events," or did we have here a case of two unwitting desperados engaged in a very timely game of political lifebelt throwing. But the "Maggie and Leo Show", just like any good West End hit, was deemed to emerge with only one star and indeed it put her straight on the Oscar-winning path to another successful General Election.

However, to get back to the war, and that's exactly where she wanted us to be, when dawn broke across a clear sky on Tuesday 25th May, every man at San Carlos, standing-to by his slit trench or on action stations afloat in the Sound, knew what was coming. It was another day of perfect flying weather. It was also Argentina's National Day, and everything pointed to another big push from the air. On the ships, the gun crews huddled deeper into their storm coats and anti-flash hoods. On land, the Paras and Marines used the first light to break open their field rations and brew the first "wet" of the day.

Brigadier Julian Thompson, Commander 3 Brigade, was at that time in charge of the landing forces because Maj.-General Jeremy

Moore was still making his way south, incommunicado on QE2. Thompson started his day as usual, taking up his post in the Ops. Room, which in fact was just one section of the dense maze of dug-outs and camouflaged tents at San Carlos Settlement which was referred to as Brigade H.Q. The General laboured over his maps and carried on with the task that had obsessed him for days; he had secured the beach-head, from the land at least, and now his objective was to saddle up, break out and advance on Port Stanley.

In essence, his plan was this: wait for the reinforcements of men, machines and supplies at that moment ploughing south through the Atlantic, and then mount a straightforward assault across East Falkland to Stanley. He could see no point at all in using precious resources to diversify the attack either to the south or the west. Stanley was the only target that made real military sense. He saw the assault in terms of a rapid series of hops as his men were lifted in the big Chinook helicopters, across the daunting terrain of the island. But all this was to change by the time the Brigadier got back into his sleeping bag long after dark on 25th May.

The air raids began as usual; the hooting of the ships' sirens from the San Carlos Water and the cry of "Air raid red" through the ships' tannoys and over the radio sets spread through the hills. At first, it looked as if the air defences – on ship and land – were in for a good day. The first wave of elderly Argentine Skyhawks was intercepted as they flew in over the sea from the north-west and were turned back after four had been shot down. Then the pattern for the next two raids changed. The first pass came, as usual, screaming down Bomb Alley, a few seconds' worth of terror, but this time, little destruction. Then most of the Skyhawks and the Mirages flying with them changed direction to fly low across the radar clutter over the hills of West Falkland, dipping off the north of the Sound, heading, it seemed, for the northern picket set up by the Navy.

It was a deadly assault on the two ships standing picket duty there, H.M.S. Coventry and her Seawolf carrying guard, H.M.S. Broadsword. Most of the bombs were aimed at Coventry, and despite brave and effective retaliation, she was quickly sunk.

As Coventry sank, leaving the northern point of the Sound unguarded, two Super Etendards, armed with the dreaded Exocet Missile, flew through the open corridor towards a large ship heading for the Sound and identified by Argentine intelligence as a high-priority target, the carrier Invincible. This was the moment the Junta had dreamed of. To sink the aircraft carrier Invincible on their National Day would restore their pride and ensure a huge boost to the fortunes of General Galtieri. But it would also hinder – if not cripple – the British effort. This was not to be, because in fact, it was the Atlantic Conveyor that was hit and sunk by the lethal Exocets.

At about this time in the early afternoon, Brigadier Thompson was summoned to H.M.S. Fearless, the flagship of the landing force, to talk over the satellite to headquarters at Northwood. Major-General Richard Trant came on the line – and within a couple of minutes he had overturned everything that Thompson had been working towards. Trant had orders; not suggestions, nor requests – orders. The demand from London was simply this: attack Goose Green and advance on Stanley.

As a former Marine and try as I have with the history books, I have never been able to discover which came first – the orders to advance or the news that Conveyor had been sunk. Either way, together they destroyed Thompson's initial strategy, because Atlantic Conveyor was carrying the big Chinook helicopters, and about 9 Wessex helicopters as well. All but one Chinook were lost, along with such essentials as tents for 4,000 men. It was the biggest logistical blow of the war, and it cost Thompson the engine-room of his strategy for getting his men across the Falklands. He was left with orders to advance for battle, with no apparent means of getting his men into an assault position. This dilemma was never seen in London. There, the political pressure was always intense to do something, get a result. Britain had already lost five ships – the politicians wanted success in return. There was increasing pressure for action before the world's most expensive talking-shop, the U.N., ordered a ceasefire leaving Britain with nothing but the beach-head.

At least the unambiguous orders resolved the dilemma, even if they replaced it with a huge risk. Thompson called an orders briefing for his unit commanders. The C.O.'s helicopters buzzed over to Port San Carlos, Sussex Mountains and Ajax Bay and then back to Brigade H.Q. There, the Para and Marine Colonels received the news with disbelief.

2 Para were to go to Goose Green – Brigade could just scrape up enough Sea King helicopters to take some guns and supplies there for them, but the fighting men would have to walk. 3 Para and 45 Commando were to advance on Stanley, via the northern route. There would be no airlift. Further, since there were no vehicles capable of carrying men and equipment across that marshy, rocky, mountainous country, it would be a matter of "yomping" on foot and of carrying enough gear to take on the enemy along the way if necessary, and this also meant rations to live on.

At this point in time in this new century it must be worth reflecting, as trains are expected to run on systems falling into ruin, and old peoples' care facilities up and down the country are closing on a daily basis for lack of funding, while new establishments spring up everywhere for economic migrants, I think it fair to assume that Julian Thompson would be forgiven for looking back in wonder at how the politics of the madhouse have never really changed in this country. No doubt some boffin at the Ministry will berate me for such foul and unprecedented utterances but look, I can take it – I have a green beret!

So to end this section on another reflective note General Thompson, under his cramped canvas at San Carlos, kept his thoughts largely to himself. They were not happy thoughts. The politicians and the top brass had forced his hand – but his was now the responsibility. His only real comfort was an enormous faith in the qualities of the men who were now struggling out under huge loads into the peat bogs and rock-runs. His faith had not been misplaced, as every truthful soldier in dire straits will confirm. The people who influence you most, are the people who believe in you and looking back, in spite of the drawbacks I would dare to say that Thompson's optimism was the light that led to real achievement. So on they went to victory, for as every old bear knows, when winter comes, can spring be far behind?

"The Argentine Demise"

Among the reasons advanced by the Argentines to explain their defeat in the Falklands was the claim that the British Infantry had more sophisticated weapons. They made unwarranted claims for British night vision and communications equipment – claims which were not only false, but ignored the fact that certain types of British equipment were inferior to their own. In fact, throughout the campaign, British Infantry were better equipped than their Argentine counterparts only in the matter of anti-tank weapons.

Both sides were equipped with virtually identical versions of the most common personal weapon, the rifle. The Argentine factory at Rosario produces F.N. F.A.L. NATO pattern rifles in three variations, and the Argentine Infantry generally used the F.N.50-61, with folding stock, although some of them preferred the F.N.50-63, which is a little lighter and has a shorter barrel. All the Argentine versions were capable of automatic fire, but the British rifles – which were F.N. F.A.A., but in the L.I.A.I. British type – had the self-loading mechanism but could not fire bursts. All weapons fired a standard 7.62 mm round, which is accurate and effective up to 875 yards. The British had deliberately rejected the option of automatic fire when they selected and developed the L.I.A.I. It was considered wasteful, and as the British Army was always proud of a high standard of musketry training, the broad emphasis was placed on single-shot accuracy. The rifles and ammunition used by the Army in the Falklands were not modern types, and a new re-equipment programme was in hand which would provide a new rifle with lighter rounds that would "tumble" on impact, inflicting far worse wounds.

Just before this military period, prompted by the needs of soldiers in the Vietnam War, a major change was happening to military rifle bullets which challenged both medical skills and the laws of war; the bullets got smaller. Jim Sullivan was one of the designers of the M-16 rifle (the Armalite as we know it), and of N.A.T.O.'s small calibre bullet. His purpose was to give soldiers more fire-power. The smaller bullet plus cartridge designed for the M-16 had many military advantages over its predecessor. If you look at 7.62 and 5.56 mm rounds side by side, you can imagine how much more controllable a gun would be that fired the smaller one. In addition to being more controllable, the smaller cartridge allows the man to carry twice as much ammunition and therefore have twice the fire-power.

But when they first appeared, the new smaller bullets had some unexpected medical effects. So, before we continue with the weapons of the Falklands period, it is a good idea to look at what was available but not, to my knowledge, used by the broader mass of infantry in that particular conflict. Professor Jim Ryan (Professor of Military Surgery, Royal Army Medical Corps, Millbank, London), was quick to bring to the fore that as bullets became smaller and lighter, from 7.62 mm down to 5.56 mm, although the rounds were still jacketed, the jackets plainly were thinner, less strong, more likely to break up or to deform or fragment. Now if this happens, from a clinical point of view what you see is a more severe wound. When a bullet hits the target, a human body for example, then it penetrates, and because it is spinning it is very stable and it penetrates about 5-15 centimetres in a stable position. Then it begins to turn. The bullet turns sideways, then backwards, then back to a sideways position. In this sideways position it runs through the body up to the last moment where it turns again into a backward position, which is how it comes to rest. At the moment when the bullet turns the first time into a sideways position, there is a big bending stress along the bullet, and it is possible that the bullet breaks up at this moment. This copper and lead fragmentation in every direction can cause massive injury.

Another weapon available and on offer to all who wanted it was the U.S. Army (Advanced Combat Rifle). The A.A.I. weapon fires a 10.2 grain steel dart called a flechette. These fin-stabilised projectiles are fired at a muzzle velocity of 4,600 feet per second. Flechettes have proven to be effective against hard targets and body armour, and have shown to deform and bend in soft targets, making them as lethal as bullets. They are another example of fragmentation injuries, small high velocity needles developed to get through many forms of modern body armour. If you look at much of today's body armour under a microscope it has a weave, and these little flechettes will get between the weave. From a surgeon's point of view, these are very worrying because the entry wounds may be very small. You can imagine that a harassed and tired doctor may have very little to see as he looks at a downed soldier. He may have to undress him fully and even then, if for instance the flechettes have gone in over the shoulders or in the back, he's going to have to look very carefully before he realises that he's faced with somebody who has multiple flechettes down perhaps through his chest and perhaps further down into the abdomen.

Because flechettes were used in Vietnam in artillery shells and in shotgun rounds, hugely effective against massed infantry, there were studies done mainly in the 1970's and at the negotiations in Geneva from 1978-80, suggesting that flechettes caused unnecessary suffering because some of them tended to bend on impact. When you consider that most small-calibre projectiles can fragment on impact with soft tissue at certain velocities, obviously the difference in wounding was not significant and the idea that flechettes caused unnecessary suffering did not prove to be the case. The claim just didn't hold water.

I think war, if we're going to have it – and it seems we always will – is going to always be a horrible affair because it involves killing people. We've just had a brief look at some battlefield weapons and different types of wounding, which is just a small part of what happens in war, but thankfully during the Falkland conflict we tried to stay – loosely speaking – very conventional as to the weapons we used, although the Argentinians had some nasty surprises up their sleeves had they been given the opportunity. In 1982, however, the advantage may have been slightly on the side of the Argentine rifles. They had six weeks to prepare their defensive positions and bring up adequate ammunition, enabling their automatic weapons to put down daunting fire as they reeled off bursts from their 20-round, quickly changed box magazines. The British were short of ammunition and had to march many miles into action carrying loaded weapons that were by modern standards cumbersome and heavy (10 _ lbs). A further drawback was that the 7.62 round lacked the range, accuracy and penetrative power of the old-fashioned and obsolete .303 cartridges, giving the riflemen less chance of winning a firefight with entrenched defenders than their fathers and great-grandfathers had in both World Wars. It is also remarkable that, despite the L.I.A.I.'s very low rate of fire, British soldiers were on occasions forced to use their bayonets.

Apart from their rifles, there were some lighter personal weapons distributed on both sides. For some of the officers (generally those of field rank and above), and for troops encumbered with heavy equipment, the British Army and other branches of the service provided the 9 mm Sterling sub-machine gun, which has a magazine capable of holding 32 rounds and can fire automatically or in single shots. This weapon provides great firepower but, like all sub-machine guns, it is inaccurate and is used principally in self defence.

The Argentines had a similar type of weapon, the PA3-DM, which had a 25-round box magazine and a performance which is roughly equal to that of the Sterling. In addition to these standard issue light weapons, the British Special Forces wielded a variety of types to suit their personal taste. Perhaps the most common was the 5.56 mm Colt "Commando" XM 177, a cut-down version of the M16 (Armalite) assault rifle. The Colt "Commando" has the hitting power of a rifle, but because it is a shortened version, it lacks accuracy at long ranges.

For the ordinary infantry on both sides, the first clear diversity in weapons occurred at the section level and above in the differing choice of mechanism. The machine gun has two functions in an infantry battalion; its first role is as a light, portable weapon, adding range and firepower to each section; its second is as a much heavier back-up weapon at support company level, where it can be deployed to assist the battalion with sustained fire up to a considerable range. For many years after World War II the British used the Bren gun as an L.M.G., and various pieces of Vickers or Browning ordnance for sustained fire. By the 1960's the British had come to the conclusion that a general-purpose machine gun (G.P.M.G.) could combine the two roles of the L.M.G. and sustained-fire (S.F.) support. As a result, the infantry battalions engaged in the Falklands Campaign were equipped throughout – from section to support company – with a British version Belgian mag, designated the L7A2 (G.P.M.G.) and firing a 7.62 mm round.

The advantage of having one M.G. instead of two lay in the fact that every soldier could be trained to handle it, and all the ammunition was interchangeable (the Vickers Mk I S.F. weapons had indeed used .303 cartridges, in common with British L.M.G.'s and rifles, but it derived its extra range from the use of Mk 8Z ammunition as opposed to Mk 7). The disadvantage lay in the G.P.M.G.'s weight – at 24 lb unloaded, heavy for a section weapon – and the feeling of many was that it was not sufficiently powerful for an S.F. role. It was also used habitually as a belt-fed weapon; there was a suspicion that ammunition belts snag in trees and shrubbery so that a box magazine was more suitable for an L.M.G. Despite these nagging doubts, the G.P.M.G. is still very popular with the men who use it and it proved its value in the Falklands War.

Experience has taught the British Army the importance of machine guns. It is one of those armies – rare in the modern world – that is constantly in action. Many of the officers who went to the Falklands had first hand experience of war, and they made sure they took with them every M.G. on which they could lay their hands. This meant that there were dozens of G.P.M.G.'s taken into each battalion attack in their bipod-supported L.M.G. role, and a fair number with each support company mounted on buffered tripods to exploit their range of 1300 yards and high rate of fire of 750-1000 rounds per minute in the S.F. role.

In addition to G.P.M.G.'s there were older M.G. types ransacked from stores by resourceful British soldiers. There were a number of Bren L.M.G.'s, that "perfect weapon of war", used as personal weapons despite their individual weight of 21 lb without a 30-round box magazine. I myself can fully understand why a soldier would choose a Bren as a "personal weapon" despite its weight, as I carried one as a section gunner with the Warwickshires in both Aden and then the New Territories of Hong Kong in the 1959/60's. If you persist with this M.G. over many years, no matter what the terrain, it becomes a sort of nonentity as far as burdensome kit is concerned. With a magazine in place, there is a method of hooking it around the back of the neck with the central mass of the gun's weight balanced on the shoulder muscles and even on steep hill-climbs – for which the New Territories is renowned – you can travel hands-free until the moment you need it and then take its weight back onto your arms again. If it had one drawback it would be for the No. 2 on the gun. His job was to stick to the No. 1 gunner at all costs and this meant lugging a box of 12 spare magazines (30 rounds in each mag.) everywhere the gunner went. It went without saying that the gunner could move faster than the No. 2 and I learned the lessons of this on my very first outings as a young No. 2 to a gunner in Aden with the Warwickshires. From our tented unit location in Aden we were on frequent forays up-country to Dhala, a terrain in which it was a feat of endurance in itself to lug boxes of loaded Bren mags. during constantly practised fire and movement drills in boiling temperatures. Then, a few weeks later to my amazement, my No. 1 handed his gun and all of its ancillaries over to me with a big demob smile on his face, informing me that "Khormaksar" – the R.A.F. base in Aden

– were flying him home the next day as he had finished his (N.S.) "National Service".

Well, for me the Bren very quickly became a personal weapon as I got to know and master it for the benefit of 3 Section – 9 Platoon-Charlie Company The Royal Warwickshire Regiment. From our stint in Aden we embarked onto the old troopship "Nevasa", a sort of mini Canberra. This led to a quiet, restful period as we left the Gulf of Aden, crossed the Indian Ocean to Ceylon (Sri Lanka), then on and eastward for a three-year stint in Hong Kong. The fragrant harbour. Apart from the favourite Bren and to add weight to the S.F. forces, some Browning heavy-barrel MZ's were pressed back into service with their 50-calibre ball ammunition and stolid 450 to 500 rounds per minute rate of fire. However, the Browning weighs 84 lb, even without a suitable tripod mounting, and it was used principally to add to anti-aircraft defences. The inclusion of so much extra automatic fire from machine guns proves that the British sensibly attached great importance to overwhelming their adversary in a fire-fight.

In contrast, the Argentines stuck to the principle of using L.M.G.'s at section level and backing them up with S.F. from heavier calibre support weapons. The L.M.G. was simply their rifle with a heavy barrel and attached bipod, designated the FN 50-41. Nit-pickers would probably insist that the FN 50-41 was a machine rifle rather than a genuine machine gun, because although the heavy barrel meant that the weapon could maintain a higher rate of fire than the automatic rifle, there was no improvement in range nor much in accuracy. While the British G.P.M.G. was and still is by far the most important source for generations of firepower in a section, the Argentine FN 50-41 gunner could not provide the much-needed stopping power for his immediate comrades. The Argentine weapon had the attraction of being very light at 13 lb unloaded, but would come second best in a performance contest with the G.P.M.G.

With such a light section weapon, the Argentines were forced to provide S.F. backup with much heavier M.G.'s - .5 inch and some .3 inch Browning M.G.'s. The .3 inch is obsolescent but still an excellent weapon, and both calibres alike in construction and performance.

The Argentines were not glaringly deficient in M.G. support, although they experienced some problems caused by the use of different ammunition types. It was in the provision of rocket-launched anti-tank projection that their ground forces were completely outmatched by the British. Accurate medium-range anti-tank weapons were distributed among Argentine infantry only at support level, while British infantry bristled with powerful rocket launchers in every section. Of course, the Argentines had less cause to use anti-tank weapons as the British maintained manoeuvrability over open ground, but they did site their 105 mm recoilless gun model 1968 to great effect – particularly in the battle for Mount Longdon. However, they must have been stunned by the liberality with which the British used anti-tank rounds against their sangers and entrenched strongpoints. The British weapons were the "Carl Gustav" 84 mm recoilless rifle which fires 6 _ lb rounds, and the U.S. made 66 mm LAW one-shot throwaway assembly package, each one of which weighs only 5 lb. The 84 mm "Charlie G" as it's called, is very accurate. With a x3 magnification telescope sight and an effective range up to 765 yards, its projectiles can burn a hole through 1 _ inches of plate armour, which gives them plenty of power to break up bunkers.

Also very effective against ice cream vans: (see 45 Commando, Chapter 20)!

The 66 mm LAW is obsolescent because it is not capable of taking on the main battle tanks of the Warsaw Pact but in this new age I guess that just doesn't matter any more. But it proved good enough to chew through Argentine infantry strongpoints and is accurate up to 165 yards. Some observers drew the conclusion that rocket launchers gave the British infantry their advantage over the Argentines. The credit, however, lies equally with the ferocious determination of the British infantrymen, who carried such a great weight of firepower in machine guns and anti-tank weapons into action, and then had the skill to use it properly. Even this firepower did not silence all the Argentine positions, and some more old-fashioned weaponry had to be used in close quarters combat. Grenades are most effectively used to winkle enemy out of entrenched or enclosed positions and, during the Falklands campaign, it was the Argentines who occupied the trenches and the British who were winkling them out. As a result, the most effective use of grenades was made by the British. In the final count the British relied on bayonets, and it is something of a curiosity that the British are still so wedded to the idea of using this ancient but deadly weapon.

Indeed, it was in the philosophy behind their use of infantry weapons that the British proved themselves so superior to the Argentines. They believed, correctly, in winning the fire-fight with a heavy volume of fire. At section level the weight of fire from their machine guns and rocket launchers proved decisive. It is worth mentioning, however, that although the Argentines seemed particularly vulnerable to night attacks, this was not because of the technical superiority of British night vision equipment. Both sides used image intensifiers, which gather in whatever light is illuminating the target (starlight or moonlight) for no night is actually pitch black, and amplifies it to contrast the target with its surroundings. The British devices used by the infantry were so-called "first-generation" equipment capable of providing their snipers and observers with much better vision.

A contrast between the equipment of the Argentine and British infantry equipment does not show any great British advantage. If there was any great difference, it lay in the quality and training methods of the men who were trusted to use it.

CHAPTER 24

"REFLECTIONS OF VALOUR"
STORIES OF THOSE WHO WERE THERE

FIGHTING PATROL

For Lt. David Stewart M.C. of X-Ray Company 45 Commando, this was going to be a night to remember.

On the afternoon of 4th June 1982, 45 Commando Group arrived at Bluff Cove Peak on the lower slopes of Mount Kent. The fourth set of defensive trenches were dug and they settled down to await their next move. They were all aware of the fact that this was the first time that they were actually within striking distance of the enemy ground forces around Port Stanley.

On the morning of 8th June, Lt. Stewart was called for an "Orders Group" by Captain Ian Gardiner, Royal Marines, his Company Commander, where he was told of the next night's activities. He was to prepare for a fighting patrol for the night of the 9th as a divisional attack on forces around Stanley had been put off to beyond 9th/10th June. He warned his troop sergeant, Sgt. Jolly, and briefed his corporals to be ready for the task. The next morning he was up at dawn and wasted no time in organising the day's activities. He was going to take his three corporals, along with Corporal Wilkie of Reconnaissance Troop, up to a saddle overlooking the Two Sisters so as to get an idea of the terrain. When he had completed his recce, which was carried out in bright sunshine, he returned down to the Commando H.Q. where he was briefed by Captain Mike Hitchcock, the Operations Officer. His mission was to "cause casualties and harass the enemy". He was also to attempt to map out as many of the enemy positions for the main attack on the night of the 11th as possible.

After this brief, Lt. Cook briefed him on his signals back-up, radio frequencies, and code words. Before he left the main H.Q. area, he had a ten minute chat with Chris Fox; ten very useful minutes. It then took him a little over an hour to write his orders, incorporating all the information and advice from the above experienced sources. It was unfortunate that the previous night's patrols carried out by two troops, one each from "Z" and "Y" Companies, had been forced back by a heavy artillery barrage.

At 14.00 hours, Lt. Stewart assembled the troop in a gully to issue his orders in bright sunshine. The Chaplain, Wynne Jones, sat in and, given half a chance, would have gone with them! Cpl. Wilkie was to lead the route in with his three men, all of whom were at one time in "X" Company and were known by many of the troop. Seven Section, under Cpl. Colville, was to be the assault section to attack positions located by the recce group. Nine Section, under Cpl. Knott, was to be the fire section to cover any attack put in by Cpl. Colville. Eight Section, under the less experienced Cpl. Tanner, was to be in reserve under the watchful eye of Sgt. Jolly. The troop H.Q. consisted of Lt. Stewart and his radio operator, along with Mne. Lepier, the company signaller. The 2-inch mortar team, Mne. Greer and Mne. Cluman, was to be retained, but the 84 mm Carl Gustav team was dispatched to one of the sections as riflemen. Cpl. Thompson of the assault engineers was to lead a two-man team for mine clearance and laying explosive charges. Bombardier Engleson from the Gun Battery was to head a two-man forward observation officer (FOO) team for artillery support.

Much organisation, responsibility and a lot to think about as the patrol moved out into the darkness to, as one man put it, "tread the black peripheries of eternity".

The emphasis had to be on stealth so as to ensure them the initiative over the enemy; an early contact could seriously jeopardise the mission. Lt. Stewart was warned to keep away from the Murrel Bridge area due to a para ambush which had been set up there, and to watch out for one of the Surveillance Sections returning from forward of the lines.

The distribution of fire-power was the strongest one could expect for a patrol like this. Each section had two automatic weapons, one G.P.M.G. and one L.M.G. and Nine Section had two L.M.G.'s. Each man carried two grenades, an even distribution of LZ's, and white phosphorus smoke grenades. Seventeen 66 mm anti-tank rockets were liberally spread throughout the troop. The inclusion of the FOO party with his artillery support on call considerably boosted their fire power.

At 15.30 hours Lt. Stewart rehearsed the troop in the order of march, code words, password, "Actions On", and the nitty-gritty of section tactics. At 16.00 hours they were ready and fully prepared. Captain Gardiner, who until then had kept his distance, came over to talk to the troop. The troop commander remembered him being impressed by the spirit of the men and, in his characteristic way, he boosted the morale even higher. All we could do now was wait for 01.00 hours, when we were to move out.

At 23.00 hours, a call from Commando H.Q. warned Lt. Stewart of lights showing on Two Sisters, which are two hills side by side, confirmation of the enemy's presence. At 00.15 hours he moved out into the cold moonlight to prepare his kit, and he could see the dark shapes of his men as they moved about, putting on their kit in preparation for the most momentous night of their lives. Half an hour before they were to leave, Cpl. Tanner reported three of the men had gone down with acute stomach pains. Lt. Stewart had always been worried of the possibility of water poisoning and, at last, the shortage of sterilising tablets had taken its toll. The men had been feeling rough all day and, until now, had said nothing, but they could hide their suffering no longer, so he had no choice but to do a quick reshuffle of weapons and leave the three behind. At 01.00 hours they moved out of the relative safety of their defended location and up to the saddle, where, some hours earlier, they had carried out the recce. From there they moved down towards the unmarked feature between Mount Kent and Mount Challenger. They moved in staggered file, well spaced due to the exceptionally good visibility, and paused on the lower slopes to have a good look at the Two Sisters feature, still some four kilometres off. The Troop Commander, Lt. Stewart was struck by the similarity of the situation to a fighting patrol he led during training as a young officer at Lympstone. He was under no illusions, however, that they were facing a small number of the training team firing blanks. But tonight it was for real. They moved on, crossing a patch of open ground, down towards the tip of a rocky outcrop some 1000 metres from the objective. As they moved over some cratered ground, they prayed that the enemy artillery would remain silent. Had it not done so, they would have had little chance. One of the ships of the Task Force was pounding Mount Harriet with Naval gunfire. It was an impressive sight and somewhat reassuring, as it may well have distracted the enemy on Two Sisters.

They reached the final piece of cover before the long open stretch to the objective at 04.00 hours, where Lt. Stewart ordered a rest and a final check of equipment. The immediate problem was how to cross this open ground tight under the enemy's nose without being detected. He eventually decided that each section should move towards the stream and over to the far bank in tight groups, each separated by 10 minute intervals. This technique had apparently been used to effect by a previous patrol in this area. The "rule book" described a number of formations, each suited to different terrain, but this ground was so open and flat that any conventional formation would have been instantly recognisable as troops. As the ground was well pitted with shell holes, one more dark hole might go undetected, if it were moving slowly.

At 04.45 hours they had all crossed the stream and were moving up the stream towards the objective. By 04.50 hours they had reached the rocks and Lt. Stewart began deploying the sections in preparation for the move up the feature. As the last section was moving into position, Mne. Crawford of the Recce Section spotted movement some 200 metres ahead. This was confirmed as an enemy by Cpl. Colville, who had also seen him with the aid of a night-viewing device. Stewart deployed Cpl. Colville up towards the enemy position, while Cpl. Knott moved out to the left, with his gun group of L/Cpl. Hind and Mne. Hayter furthest out to cover the assault. Sgt. Jolly and the Reserve Section were to the right rear and H.Q. in the centre of the rocks. The time was 05.00 hours. Stewart passed word to Cpl. Colville and Cpl. Wilkie to move up towards the enemy location, while he waited with his mortar team. At 05.00 hours Cpl. Wilkie radioed to Stewart that he could see two men with a machine gun and that he was going forward. A few seconds later Stewart could hear rifle fire from Wilkie's position. Someone shouted for some light and the mortar instantly fired two well-aimed illumination rounds. Wilkie related later that, as they had gone forward, a third Argentine had rounded the rocks and both parties were in direct confrontation. In the ensuing fire fight, the three Argentines were killed and Cpl. Colville moved round under cover of his gun group to continue the advance.

Meanwhile, Stewart had heard: "Two dead, moving forward," over the radio. At the same time, Cpl. Knott shouted that he could see figures running down the left side. He leapt over to Knott and could see from their position that they had to be enemy, so gave the order to open fire. Four more enemy were killed. He then gave the order for his gun group to stay put and cover the rest of the section up the left side. Moving to the front of the section they then made their way among the rocks but, as they were about to round a large boulder, a streak of tracer came down a few feet to their left. It was evident that Argentine positions further up the slope had been alerted and they now poured down fire from above the Marines, forcing them back to the shelter of the rocks. The 66 mm rocket was working extremely well, but they really could have done with some high explosive rounds for the mortar! When Lt. Stewart arrived back, he found that Cpl. Colville had also been forced back by the same fire and he decided that they could serve no further useful purpose by staying on that feature.

He then called Bombardier Engleson to bring down fire as soon as they were clear. Cpl. Knott's ammunition was getting low and Cpl. Colville had used up about one third, but they had no casualties. They knew that moving down the exposed slope was going to be a problem as it was, and a lone machine gun out to their right was putting down fire directly across their rear. Stewart despatched Mne. Marshall up, with his unit, and Sgt. Jolly to silence it. They moved up onto a rock feature, not the safest of places under the circumstances, and, with a few bursts of well-aimed fire, the enemy location was neutralised. He then ordered each section to turn to a hollow some 100 metres back. When they were all well tucked down, Stewart could identify more easily the enemy position. The ground behind them was being well covered by fire from the enemy, and they were also using some sort of mortar. The artillery support was not yet ranged onto the Argentine positions, so Stewart ordered a fire-fight. During this phase, a number of 66 mm were fired, Mne. Fletcher proving his proficiency with the weapon. After some minutes (hours!) each rifle group was despatched back under the cover of the remaining gun groups, and each gun group was then in turn ordered back. Lt. Stewart remembered looking back and seeing rounds hitting the ground all around the running men. Up till then he had been too occupied to be frightened; now, however, he was worried because a casualty at this stage on that terrain would have been a disaster of nightmarish proportions.

Sgt. Jolly and Cpl. Thompson were running five metres apart, before being separated by a line of tracer. Anyone who makes the mistake of thinking that Assault Engineers are not fit should challenge Cpl. Thompson! Stewart had no idea how long it took to reach the stream but, as they reached and crossed it, the small arms fire was falling well behind them. He was thankful to be able to slow the pace now, as they made their way back to the ridge opposite Two Sisters. It was not over yet, however. As they reached the rocks, enemy artillery ranged down on them, so they made a bee-line for the far side of the rocks. When he was sure that they were safe, he called a halt and they attempted to recover their breath. Their ammunition was very low by now, but they still had no casualties. Someone was smiling down on them. Lt. Stewart's first contact report had been sent at 05.10 hours, but he now gave a more detailed description of events and informed Commando H.Q. of his intention to move back. To his surprise, he found himself talking to his Company Commander; he had been in the watch-keepers BV all night.

They finally stumbled back into Bluff Cove Peak at 09.00 hours and Lt. Stewart went away to the Intelligence Officer for debriefing. He then, at last, "crashed out" at about 12.00 hours. Lt. David James Stewart was awarded the Military Cross.

CHAPTER 25

THE TAKING OF TWO SISTERS

"Give me 120 men and I could die of old age defending this hill."

Lt. Col. A. F. Whitehead D.S.O., R.M., C.O. 45 C.D.O.

It was daylight and Lieutenant Chris Fox, together with his recce patrol, was lying up on the enemy slopes of Two Sisters. They were efficiently concealed, as well they might be; it was several miles back to their assault base across a wide open valley and over the ridge to Mount Kent.

They had moved forward to this dangerous position just 24 hours after completing, with the rest of 45 Commando, their epic yomp from San Carlos. Their mission was to locate Argentine bunkers, and now they were waiting for darkness to fall before returning to base. Darkness was the only truly effective cover on those bare Falkland hillsides.

An Argentine patrol literally stumbled on them. Rifle fire poured down without warning from above, most of it frighteningly accurate. Fox received a bullet through the hand. He had no orders to fight back under those circumstances; on the other hand he had no choice. He and his men returned a devastating hail of S.L.R. and sub-machine gun fire. Twelve Argentines were killed, and a further three suspected hit. When darkness arrived, the Marines slipped away without casualties. It was an interesting start to 45 Commando's battle for Two Sisters; almost as many Argentines died in the clash as in the main assault which followed later. Lt. Fox was flown back to the medical centre at Ajax Bay, but soon rejoined his unit with a stitched hand and scarred finger.

No-one knew at that stage when the real attack on Two Sisters would take place; most of the men had been expecting to go into the assault within 24 hours. As Colour Sergeant Bill Eades had said on the march, "There's only one way out of this and that's through Stanley," and now Stanley was tantalisingly close beyond the twin rocky peaks of Two Sisters. You could see the airport and the harbour from the ridge above the Commando base on Mount Kent. But in the end they had to wait eight days for 5 Infantry Brigade to arrive as reinforcement, and for Brigadier Thompson to decide to give the order to go. Learning the lesson of the under-supported attack at Goose Green, he had decreed a maximum level of artillery support for the battles in the hills around Stanley. Whenever the mist and blizzard let up, the helicopters would come droning up the valleys with huge pallets of ammunition slung in nets under their bellies. Eventually five batteries were dug in around Mount Kent, each with six 105 mm guns, every gun with 1200 rounds.

Two nights before the attack was finally called, 45 was hit by their worst disaster of the war a "blue on blue" or "friendly on friendly" clash between their own mortar troop and a patrol. In the dark the patrol was taken for Argentine; the mortars opened up, to be met with a withering return of fire. In the confusion, five died, including the mortar troop sergeant, and two were wounded. It was one of those tragedies that are inevitable in war, but while 45 took it philosophically enough, everyone felt it a blemish on their exemplary record in the Falklands. If anything, the tragedy increased the determination of the C.O., Lt.-Col. Whitehead, to take Two Sisters with the minimum casualties. He had been critical of the level of casualties at Goose Green and said before his own assault, "Let's finish what promised to be the last lap in style. I want no futile and useless casualties."

The battle plan for the night of Friday 11th June was described by one officer as a "straight forward, Warminster-style assault". By this he meant the orthodox infantry attack plan taught since World War II. Although the approach to the enemy lines would be "silent", the battle itself would be "noisy" with the maximum weight of artillery hurled at the Argentines as the riflemen stormed their positions. While 45 Commando would take Two Sisters, 3 Para would assault

Mount Longdon to the north and 42 Commando Mount Harriet to the south. H.M.S. Glamorgan was to provide Naval gunfire support.

"Late Start to the Battle"

The start line was at Murrell Bridge, a rickety structure over the Murrell River at the foot of Mount Kent. The lead rifle company for the assault, "X" Company, led by the able and articulate Captain Ian Gardiner, should have been there at midnight. But the Commandos were two hours late at the start line, bogged down by peat and slowed by fearsome rock-runs over which they had had to hump the heavy weapons. Whitehead was on the radio net threatening to "come down and kick" the rifle company into action as they finally began to cross the dead ground towards the southern ridge of the hill. Capt. Gardiner's company made the lower ridge of the southern peak of the Two Sisters without problems but were then promptly pinned down by ferocious machine gun fire, followed by raining mortar bombs.

A night battle on this scale is an extraordinary experience. In my experience I think it is fair to say that most young Marines expected some sort of visual impact, with men running, guns firing, targets falling and so on. In fact it is nothing like that. Noise is the predominant impact; the whoosh and thunder of incoming artillery, the enormous bangs from the British guns behind, the steady pok-pok of the Argentintines' 0.5 inch heavy machine guns sending down a stream of red tracer from the bunkers above. As "X-Ray" scratched what cover they could from the rocks and peat ridges, trying to pin-point the enemy bunkers and working out routes to their trenches, Whitehead coolly decided to change the whole battle plan. It was clear that there was no way "X-Ray" could storm the twin peaks without massive casualties, so he decided to bring in the other two companies to advance up the northern peak. In a series of quiet commands typical of the Colonel's style, he repositioned his men for a steady, stealthy assault up the hillsides each side of the main saddle between the peaks. As the rifle companies found what shelter they could on the peaks, enemy artillery still pounded in, sending up sprays of peat from the saddle between them. The Marines were carrying no sleeping bags or shelters. Exhausted, they began to slump down in the snow drifting between the rocks. Whitehead, anxious for their welfare, ordered them to find what kit they could among the Argentines' positions, and as the morning drew on, his men could be seen setting up an instant camp from salvaged ponchos and sleeping bags.

As the helicopters crept along the contours of the hill to pick up dead and wounded, seemingly oblivious as usual to fog or enemy fire, Whitehead gave his characteristically curt verdict on the fortifications he had just stormed: "Give me 120 men and I could die of old age defending this hill."

For the men of 45, the battle of Two Sisters had been no pushover, however. Whitehead's inspired planning had kept the level of casualties to a minimum, but the Paras on Mount Longdon hadn't seen it that way; the sheer weight of fire coming down on the Marines had excited the sympathy even of the Paras. By the same token, the Marines watched with awe the firefight on Mount Longdon, one man saying, "Not even the Paras'll be able to walk on water after this one." A young Marine, asked to write a minimum of ten words about the battle, had only this to say: "It was a cold and dark night the time we took Two Sisters. I am still trying to forget that night, so I will write no more about it."

Others dug in and waited for the next battle, numbed by what they had experienced. They had all been professional before, but now they were truly battle-hardened veterans. Sadly, and unknown to the Marines, the gun support frigate Glamorgan escaped the battle less lightly than they. Just as her commander, Captain Mike Barrow, was shutting down the guns to make the customary run to sea from the gun line, away from the day-time air threat, officers on watch reported a flash of light heading towards them from the Stanley shore line. The missile ploughed into the rear superstructure, exploding in the helicopter hangar. It killed 13 men, and the fierce fire that broke out injured some 20 more. However, the fire was controlled, and she steamed away to make repairs and fight again another day.

"X-Ray" Company in 45 Commando has always had special memories in my mind as, despite the drafting policy of the Core, there was a period of time from 1969 to 1978 when I was with "X" for seven years from Stonehouse Bks. To Condor in Arbroath without being drafted. When they finally caught on to me and gave me a warning against becoming part of the fixtures and fittings of "X" Company, all they did was send a few signals or phone calls here and there, shuffle me across the parade ground and straight into the Mountain and Arctic Warfare Cadre. So my "draft" was no more than two hundred yards which, as they say, "did me just fine." However, let us move on to some more Falkland stories which deserve airing as reflections of valour.

MOUNT HARRIET: STEEP AND LONELY

This attack took place on the night of 11th June and Lt. Col. Nick Vaux took a route that was least expected by the Argentine troops on Harriet. He attacked from the east, after taking the Commando through a minefield to the south and making a 180-degree turn to approach the feature from the least expected axis.

Harriet is a rugged outcrop of limestone due south of Two Sisters. Between them the features of Harriet and Two Sisters offer a natural screen for Mount Longdon, Tumbledown and William further to the east. Harriet and Two Sisters, as well as Mount Challenger to the west, are linked by saddles of high ground that are not only exposed by day, but are so featureless that they make navigation very difficult by night. Between Challenger and Harriet is Goat Ridge.

Lt. Col. Nick Vaux, who as a young Troop Commander had landed at Suez, had been in the area since elements of 42 Commando had landed at Mount Kent. The good visibility had allowed him to assess the positions on Harriet and prepare his plan. However, before he could proceed with detailed planning a more complete picture of enemy positions, and particularly minefields, had to be built up. "K" and "L" Companies sent out patrols and tragically two Marines in "L" lost legs when they triggered anti-personnel mines. "K" Company, tasked with the route for the attack, also put out fighting patrols to inhibit the enemy. While men were prodding the minefields, the fighting patrols ensured that the Argentine garrison remained in their bunkers. And Sergeant "Jumper" Collins led two patrols to find paths through the minefields and a covered approach south of Mount Harriet.

A fighting patrol by 1 Troop worked its way to within 20 yards of enemy positions on Harriet, when they came under fire. But their reply – 66 mm. and 84 mm. anti-tank weapons and gunfire directed by Captain Chris Romberg – enabled them to withdraw without loss, after killing six Argentines. Shell fire from Argentine 105 mm. and

155 mm. guns, and bitterly cold weather, made life more unpleasant for the men of 42 as they readied themselves for the next phase of operations. Trench foot had already affected a number of the men. Rations were generally the dehydrated Arctic type. Everyone had been warned about the liver fluke in the Falklands water and it took time to collect and boil or sterilise water – a chore performed grudgingly by men who might have been out for 16 hours in the night probing enemy positions.

"The Vaux Orders Group"

Lt. Col. Vaux quietly told his men: "Surprise and absolute silence are vital. If necessary, you must go through the old basic training business of making every man jump up and down before he starts, to check that nothing rattled. Persistent coughers must be left behind. If you find yourself in a minefield remember that you must go on. Men must not stop for their colleagues, however great the temptation. They must go through and finish the attack, or it will cost more lives in the end…… The enemy are well dug-in in very strong positions. But I believe that, once we get in among them, they will crack pretty quickly."

The plan for 42 Cdo. was to secure Harriet with a night attack in which "K" Company would outflank the enemy with an assault from the rear, before "L" exploited this advantage to overrun forward positions. "J" Company would secure the start line before moving forward to consolidate on the captured feature.

The approach march along the south of the feature was going to be a nerve-wracking phase – the ground was completely open and an alert sentry on Harriet could bring down fire on the Royal Marines. The Reconnaissance Platoon of the Welsh Guards were tasked with securing the start line with "J" Company, and there was a maddening delay of one hour as the guardsmen failed to link up in the dark. Once they were in position "K" Company moved off. During the move Lt. Col. Vaux kept the men well spaced out as they weaved through the minefield. They had reached a point 100 yards from enemy positions before they were detected. Then the fight began. The tactics used were similar to those employed in house-clearing with grenades, 66 mm. and 84 mm. Carl Gustav all coming into play. And

Naval gunfire from H.M.S. Yarmouth, 105 mm. Light Guns and 81 mm. mortars hit positions in front of the men as they advanced.

During the night the Commando had four batteries on call to give a "full regimental shoot" – one of the heaviest artillery concentrations since World War II. The C.O. of 29 Commando Rgt. R.A. co-ordinating fire support for the night actions had 'on call' a list of 47 targets. During the night his guns fired 3000 rounds, some shells falling only 50 yards from friendly forces.

When the Commando used their 66 mm. or 84 mm. anti-tank weapons they would shout, "Sixty six" or "Eighty four" as a warning to any men who might be caught in the back blast. Captain Ian MacNeill recalled the effect of these weapons; the demoralising effect of even near misses must have been considerable – most of the members of "K" and "L" Companies closest to the explosions testified that they, too, speedily got their heads down, such was the impact of the missiles exploding. The men of "K" Company were later to recall Corporal Newland who appeared to be relaxing against a rock pulling on a cigarette – he had, in fact, been shot through both legs, but continued to man the radio and command his section.

L/Cpl. Koleszar had the surprising experience of finding that two "dead" Argentine soldiers, whose boots he was trying to remove, were very much alive as they jumped up to surrender.

As the men reached the summit, an enemy hut caught fire and gave the Argentine gunners a good aiming point. Shell-fire wounded members of the Company H.Q. including the second in command. It also interrupted the fire-fights as each side dived for cover. But with first light, the Sergeant Major of "K" Company collected nearly 70 prisoners. "L" Company had to assault uphill through the rocks. Though artillery fire forced them to take cover, they realised that the real threat was from the .50 in. machine guns and groups of snipers with excellent German bolt-action weapons. Many of the bunkers around Goat Ridge had been pinpointed by patrols of the Royal Marines Mountain and Arctic Warfare Cadre and so Lt. Col. Vaux was able to use Milan anti-tank missiles against them. He admitted it was expensive – they cost about £10,000 each – "But it was our job to get rid of enemy positions." By first light the men had overrun their objectives with two officers and five Marines wounded.

"J" Company were ready to move off on the very cold morning of 12th June. All went well until they reached the slopes of Harriet where enemy defensive fire from artillery was still falling. But it ceased before the Marines were on the upper slopes.

With dawn "J" Company began to sweep through the positions to clear the remaining enemy it seemed happy to surrender. Within three hours, 58 prisoners had been processed and despatched for Company H.Q. For the men of "J" Company, the victory at Harriet was a more personal triumph since they were made up of men who had been in Naval Party 8901 – the Royal Marines who had defended Government House back in April when the Argentine force landed – and Major Mike Norman their erstwhile O.C.

Harriet yielded over 300 prisoners as well as documents that were of great value to the intelligence staff at 3 Brigade. Elsewhere on the feature among the well-built bunkers there was a litter of .50 in. machine guns, rockets and ammunition amongst which were some hollow point 9 mm. rounds – or "dum dum" bullets. In the simple shelters the Marines found foam rubber mattresses and ration packs that compared well with British G.S. or Arctic Packs. Argentine soldiers received a pack which included powdered fruit juice, beef pate, and soap and razors. There was even a small bottle of whisky with the doubtful name of "Breeders Choice". Finally, in an unopened crate, a battle-field radar set was discovered near the summit. A dozen Argentine prisoners collected at the foot of Harriet were told by sign language that they should walk, not run, otherwise they would be shot. They misunderstood and thought that they were being ordered to run not walk so that they could be shot.

The message was got across to them eventually; their war was finally over.

BATTLE FOR TOP MALO HOUSE

For the Mountain and Arctic Warfare Cadre of the Royal Marines, the Falklands war was a complete vindication of their training. The following are the words of Sergeant Derek Wilson as he tells the fascinating story of a straight fight with an equal number of the enemy's special forces:

"I was operating with the M. & A.W. Cadre in the Falklands Campaign. We inserted on the first day of the landings and operated throughout the deployment in 4-man teams. About 12 days into the tour one of our O.P.'s picked up on a group of Argentines occupying a building called Top Malo House. The boss, who at the time was Captain Rod Boswell, went to Brigade H.Q. to try and get an airstrike on it but it was too late at night to do so. He came back and we hatched a plan for the rest of the Cadre sections, which totalled about 19 men at the time, to go and take the building out.

We were supposed to do a first light attack on the building but, as it turned out, the attack went in at about one o'clock in the afternoon. This was only about two hours after first light, actually, as we were working on G.M.T. and not the local time.

The morning started off a bit dull, a bit clagged in, snowing, and we thought it would be really good for a dawn attack. As it turned out, later in the morning it became a beautiful day, blue skies, sunshine, a bit of snow on the ground. We left the LZ at about 12.30 hours and flew out with the helicopter driver. "Driver" was the right word; he really stuck to the contours. It was amazing to see how low he could get this thing. We landed about a 'K' and a half (1.5 km.) off Top Malo House itself in a re-entrant in a bog. We all jumped out of the helicopter up to our knees in the bog – it was semi-frozen at the time but we still managed to get through it. We moved from there over really rough country for about another 'K' and a half into the area of Top Malo House. We divided into two sections, a fire section and an assault group. In the fire section there was an L42 sniper rifle, two S.L.R.'s, three Armalites, six men altogether. There were two M79's as well, and they used eight 66 mm.'s to take out the building. The Armalite, L42 and S.L.R. were the weapons predominantly used, but the Armalite just didn't have the stopping power – it's a 5.56 mm. high velocity round, and it would go straight through people as opposed to knocking them down. The 7.62 mm. S.L.R.'s and L42's were really knocking them down.

From the final rendezvous, the fire group went in on the left flank while we went in to do a right angled assault. The signal for the attack to take place was a green mini-flare fixed by the boss. The fire group then fired four 66 mm.'s into the building, opened up with automatic weapons, and fired another four 66's in. At this point the assault group fired another two 66's into the building along with M79 grenade launchers and more automatic fire. There were two

M79's in the fire group and two in the assault group so the firepower was quite devastating.

We assaulted in text-book fashion, the fire group firing in as we assaulted at right angles. The enemy came out of the house, and they seemed to be very well prepared – 95 per cent had equipment on and all were carrying weapons, all had boots and jackets on. And they came out shooting, so at this point we had two guys hit, one in the upper chest, one in the bicep.

The attack seemed to go in very quickly, all we seemed to be doing was running forward, diving for cover, reloading, up again, firing, running for cover – we seemed to be doing this all the time. I couldn't see any of the enemy because all I was doing was concentrating on lobbing M79 rounds into the building. We swept round, cleared the position, and by that time the Argentines had surrendered under our superior firepower. We reorganised in a position of all-round defence and made our ammunition and casualty reports before putting out sentries and clearing up.

I was section medic at the time and I was tasked to assist with Sergeant Chris Stone who was hit in the upper chest. We though it was shrapnel initially, but it turned out to be a 7.62 armour-piercing round – it went straight through his chest and shattered his collar bone; one entry hole, two exit holes. It skipped across the top of his lung and appeared that it had done some damage elsewhere. He was in quite a lot of pain at the time, so I gave him a shot of morphine.

Once he was settled, I attended to Terry Doyle who was shot in the upper arm. He turned out to be the worst casualty because, even much later, he still had problems with that arm. It took out his bicep, broke his arm and one of the main arteries, and did quite a lot of damage to him.

Once I'd sorted those two out I went on sentry duty again. I slung the M79 over my shoulder, unslung my M16, cocked it and set the safety catch to automatic. I had my finger on the trigger so that if I was hit I'd let off a burst of automatic fire to warn the others.

Sergeant Des Boswell put his section to getting the casualties out, and we evacuated from there to Teal Inlet. The Argentine Commander was dead, he was killed by M79 grenades, though there wasn't a mark on him. I heard, although I didn't see it myself, that his second-in-command came out and said, "That was fantastic, really well done!" He thought it was a company who had done them in, although in fact we were only 19 men. It was 19 against 16, so it was almost a one-to-one battle. It was really quite tight and we were, I should say, very lucky – or quite lucky, anyway – to get away with it."

The above was recorded by Sgt. Derek Wilson but I would like to add the real value of the Top Malo House raid wasn't just the destruction of an Argentine defensive position, but the gathering of intelligence. Like any Special Forces Unit, the M. & A.W. Cadre was frequently used to snatch prisoners.

The assault on Top Malo House has been recorded in a painting by Michael Turner.

IRON RING

Less than two weeks after the landings at San Carlos the British forces had achieved a dominant position on the islands. The main Argentine force, which had remained in Port Stanley, was now surrounded. The Task Force was pushing towards the end game. It was 10th June 1982 and huge tracts of land to the west of Stanley were in the grip of two British brigades; the sea to the south, east and north was dominated by the Royal Navy. Overhead, Harriers and helicopters were able to fly unmolested (except for anti-aircraft artillery fire) to deliver their loads of rockets and bombs. General Menendez of the Argentine forces had permitted the initiative to pass to the British Amphibious Task Force.

Frigates and destroyers of the Royal Navy delivered gunfire support to the troops on the ground and bombarded shore targets within the Argentine pocket. The ships used for these tasks were mainly the County Class destroyers Glamorgan and Antrim, the type 21 and the modified type 12 frigates. The type 42 destroyers Cardiff, Glasgow and Exeter helped too, when they could be spared from air defence duties.

The number of shells fired by the ships was phenomenal. Glamorgan fired 1245, Avenger 1075, Yarmouth 1441, Arrow 902 and Active 633 of their 4.5 in. main armament in Naval Gunfire Support (N.G.S.) to the troops. The presence of the ships was a vital factor as the weight of fire of a naval gun on a R.N. warship is far greater than the weight of fire of a Royal Artillery Battery (six barrels) of 105 mm. light guns. Naval Gunfire Support was frighteningly effective.

There were five Naval Gunfire Support Forward Observer (N.G.S.F.O.) teams with 148 Forward Observation Battery of 29 Commando Regiment Royal Artillery, and these were divided among the land forces on the basis of one team each for 3 Commando Brigade and 5 Infantry Brigade – the remaining three teams being deployed with the Special Forces Patrols. The N.G.S.F.O. parties were able to call an awesome weight of fire on to Argentine positions and key points with considerable accuracy and devastating effect. As the British Land Forces put in their attacks on the various hill lines, creeping closer to Port Stanley, each battalion-sized major unit had one of the R.N. ships dedicated to it to give N.G.S. on a particular objective. The ships were able to come and go at will, and were largely unmolested by the Argentines. Only one hit was scored by Argentine fire from the land when an MM 38 Exocet missile, which had been dismounted from a ship and fitted to a towed flatbed trailer, struck the rear of H.M.S. Glamorgan. This caused loss of life but did not put the ship out of action. One reason for this was that the warhead failed to explode – typical of the Argentine lack of thoroughness.

With the seizure of the hills, the British were able to observe the Argentine positions and their gunlines. This gave the men of the Royal Artillery about a week before the fall of Port Stanley during which to register all the observation targets. As the enemy's perimeter shrank, it became easier to determine where his locations were. Harrier GR3's were used for air strikes against Argentine positions, being called in and directed on to the targets by the T.A.C.P.'s with the R.M. Commandos and by Forward Air Controllers (F.A.C.'s) marching with the troops. Helicopters were also used in a gunship role to attack the enemy, often using G.P.M.G.-S.F.'s, manned by the crewmen firing through the right-hand door port, and 68 mm. SNEB rockets mounted in pods fitted to the outboard pylons.

General Menendez also had to contend with the possibility that Task Force warships would try to force a passage into Port Stanley Harbour or land troops in Argentine-held territory. The British had enough troops available to put ashore parties of up to battalion strength. To guard against the threat – especially the possibility of

Port Stanley being taken out – Menendez had to deploy troops to guard the vulnerable coasts and various key points to the east and south of Port Stanley. This limited the number of Argentine soldiers and Marines available to face the main British advance.

"Defending the Vital Ground"

It was clear that vital ground, without control of which General Menendez would find his position becoming untenable, was the series of hills close in to Port Stanley. Menendez had to hold Sapper Hill, Mount William, Mount Tumbledown, Mount Longdon and Wireless Ridge at all costs. Farther out from these, and constituting "ground of tactical importance" (without the control of which his position would become difficult to defend) were Mount Harriet and Two Sisters. Farther out still to the west were the key features of the Murrel Heights, Mount Challenger and Mount Kent. These hills would allow a potentially successful attack to be launched on Two Sisters and Mount Harriet and later on the hills of the vital ground. Possession of Mount Challenger by the British would allow the settlements of Fitzroy and Bluff Cove to be used as re-supply bases – either Forward Brigade Maintenance Areas (F.B.M.A.'s) or Distribution Points (D.P.'s) – which would greatly simplify an attack.

The British approached Port Stanley with a two-pronged advance, or pincer movement. The northern route was taken by 3 Commando Brigade, and the southern one was the line of march of 5 Infantry Brigade. Patrols from G Squadron 22 S.A.S. Regiment were able to report that Mount Kent was only lightly defended during the last days of May 1982, and D Squadron was helicoptered forward to take the hill, which they held until relieved by K Company 42 Commando some five days later. The Royal Marines were rapidly reinforced by two 81 mm. mortars and three 105 mm. light guns. The rest of 42 Commando were soon moved up to reinforce K Company, and consolidated themselves in possession of Mount Kent and Challenger to its south.

On 30th May, 45 Cdo. had reached Douglas Settlement and 3 Para were in Teal Inlet Settlement. The Paras then moved on to Estancia House, where they had a stiff battle with the Argentine defenders, and then they consolidated on the high ground. The C.V.R.T.'s of the Blues and Royals moved southwards from Teal to Kent and Challenger to give armoured and direct fire support. It became a priority to move forward the three field batteries (7, 8 and 79 Batteries) of 29 Cdo. Regt. R.A. with 1000 rounds per gun. This was a task which strained and reduced helilift capability of the Task Force after the loss of Atlantic Conveyors Chinook and Wessex helicopters. Most of the movement of the gunners was by air, while 5 Inf. Bde. moved by sea.

A telephone call by Major John Crossland of 2 Para to Reg Binney, the Farm Manager at Fitzroy, on 3rd June had established that the Argentines had pulled out the day before. Using the sole remaining Chinook, 2 Para were quickly flown forward to secure the settlement, thus saving 5 Bde. a time-consuming advance to contact. On 4th June, the Scots Guards were shipped round the southern coast on board L.S.L. Sir Tristram and the L.P.D. Intrepid to Bluff Cove. A similar move on 8th June resulted tragically in the casualties to the Welsh Guards at Bluff Cove. But the gamble meant that both the brigades were holding basically the same north-south line.

"The Final Objective"

The plan was for a three-phase divisional advance towards Port Stanley. In Phase One, 3 Cdo. Bde. were to take Mount Longdon, Mount Harriet and Two Sisters in the early hours of the morning of 12th June. The tasking was for 3 Para (with H.M.S. Avenger in direct support) to assault and capture Mount Longdon. To their south, 45 Cdo. (with H.M.S. Glamorgan in direct support) was to take Two Sisters, and farther south again the Welsh Guards (one company with two companies of 40 Cdo. under command) were to secure a start line from which 42 Cdo. (with H.M.S. Yarmouth in direct support) would assault and capture Mount Harriet. The Welsh Guards – 40 Cdo. composite battalion would remain in reserve. D and G Squadrons 22 S.A.S. Regiment, with H.M.S. Arrow in direct support, were to launch an attack on the Murrel Heights in the area of Murrel Bridge. Phase Two was to take place on the night of 12th-13th June, and was to be carried out by 5 Inf. Bde. The task was for 2 Para (with a troop of the Blues and Royals attached) to take Wireless Ridge after passing through 3 Para; 2 SG (with H.M.S. Yarmouth and Phoebe in on-call direct support, and 4 FD Regt. R.A. and 7 Battery of 29 Cdo. Regt. R.A. at priority call) to capture Mount Tumbledown; 1/7

DEOGR (with the same gunfire support) to assault Mount William; and the IWG-40 Cdo. composite battalion to take Sapper Hill. When these objectives were secured, 3 Cdo. Bde. in Phase Three would assault the Argentine positions in and around Port Stanley itself by immediately passing through the 5 Inf. Bde. objectives. In the event, Phase Two was delayed for 24 hours – to the night of 13th-14th June – and Phase Three was rendered unnecessary by the Argentines' surrender. Phase One of the divisional plan was carried out as planned. There was heavy resistance. On Mount Longdon, 3 Para fought a fierce battle with the Argentine 7 Inf. Regt., who had dug themselves in among the crags and used their snipers (equipped with second-generation image intensifying sights) very effectively. The capture of the feature cost 3 Para 23 dead and 47 wounded. One of these men was Sergeant Ian John McKay, a platoon sergeant in B Company, 3 Para. On the night of 11th/12th June, 3 Para, as discussed above, were tasked with attacking and capturing Mount Longdon, using a silent night attack. The B Company plan was for 6 Platoon to clear the southern slopes of the mountain, 5 Platoon to clear the northern slopes with 4 Platoon.

When the battle began in earnest, 5 Platoon came under heavy enemy fire and took several casualties. The platoon "went firm" on their position to reorganise and recover casualties, and 4 Platoon advancing up the northern slope, were soon stopped as well.

The Paras knew that they were up against a strong, well entrenched enemy, and the weight of fire that came their way threatened to bog down the advance. It was discovered later that an enemy company, armed with at least two 7.62 machine guns and one 0.5 inch Browning machine gun were holding the target.

Lt. Bickerdike, the 4 Platoon Commander, decided to carry out a quick recce of the enemy positions. Taking Sgt. McKay and a few others with him, he edged forward to have a look for Argentine defences. He was spotted almost immediately, and both he and his signaller were hit. McKay took over the group and, gathering another section, decided the machine guns had to be taken out immediately. The immediate target – the 0.5 inch machine gun – was built into a substantial sanger and protected by at least a section of riflemen, all of whom were firing from well-sited trenches.

The four men who went with McKay were all killed or wounded in the attack, but McKay carried on alone. In that particular "red mist" of battle he must have decided there was no going back. He was certainly of the ilk who fully realised that bending under terrible fear and pressure was no escape route into quitting. Someone once said:

"In all the trade of war, no feat is nobler than a brave retreat."

But Ian McKay was a firm believer in what his job was all about that night and also in the people behind him who depended on his performance and I believe that was the main spur that drove him forward, for he must have known that lost time would never be found again and could only be made up perhaps with even more casualties. So he drove forward. As he cleared the enemy trenches, he was hit by a sniper and fell over one of the enemy sangers. The enemy fire was substantially reduced, and this allowed the two beleaguered platoons to mount another attack on the machine gun later on. McKay's final charge eventually opened the way for the last, victorious, attack on the machine gun post and the summit. For his personal courage Sgt. McKay was awarded the Victoria Cross.

Meanwhile, on Two Sisters 45 Cdo. attacked a reinforced company of 4 Inf. Regt., who had 0.5 inch calibre machine guns in strong positions; the Commando lost four killed and eight wounded. The rest of 4 Inf. Regt. were on Mount Harriet. The Welsh Guards Recce Platoon, under Lt. Willy Sims, put in a diversionary attack on the west side of the feature using Milan A.T.G.W.'s, while 42 Cdo. came round the south and assaulted it from the rear. This classic infantry tactic took the Argentines by surprise and 42 Cdo. were able to take the feature with one killed and 13 wounded.

Thus on the morning of 12th June the British were in possession of General Menendez's Vital Ground, and poised to take the next step. He was truly surrounded by an iron ring, from which he could not escape, and his military defeat was but a matter of time.

MEMORIES OF THOSE WHO WERE THERE

"A Perfect Killing Ground"

"Few of the men who fought on Mount Harriet or Two Sisters had been in action before. They only had their training, their discipline and their unit pride to sustain them during a time which those who experienced it will never forget. The enemy strength they weren't quite sure about. They did say that the enemy would be a mixture of regulars and conscripts, with some of their Marines as well. So this was going to be our first Marine v. Marine contact, and that meant it was going to be a bit harder."

"It was a strange feeling the few hours before the attack. I was trying to gather my own thoughts. I hadn't been in this situation before. You couldn't help wondering if you were still going to be there at the same time tomorrow. But I'm sure no-one had any doubts in their minds."

"As we approached Two Sisters for the first time we saw Stanley. All of a sudden it seemed we were almost there. Our bergens were taken from us and we were told we wouldn't see them again until Stanley, so we knew we were on the final leg. Our objective was to go over the top of the feature and sweep round. It was a very quiet, moonlit night; slightly unnerving, because our route up the feature was very open – 'a perfect killing ground'. But you were there and you knew you had to do it. When we finally got the word to move out from the F.U.P. (forming up point), we went off at a slow pace up the slopes of Two Sisters. We could actually see the enemy up there. We had virtually reached the top when an illumination round went up. We went straight to ground, and, as we got to our feet when the light died, someone shouted 'Get down!' We all hit the deck and at that moment the Argies opened up. It was the first time in my life I'd seen anything like it! A few blokes returned fire but they were told to stop while Zulu Company won the firefight. It was kind of frustrating being down there and watching someone else having to do it. We were lying in a dip with streaking orange tracer going over our heads. We were quite lucky, but the guys behind us were getting it all.

One bloke cried out, moaned he'd been hit, and a medic rushed over. A minute or so later he told us the guy was dead. We weren't frightened any more. We wanted to hit back."

"We got the order to pull back a bit. With the rounds going over our heads it was a very hard thing to get up, to leave your nice little safe dip in the ground. Then we came under mortar fire and for the first ten minutes it was petrifying. You just lay there thinking, 'When's mine coming?' Luckily, it didn't."

"Then we got the order to move up. I was in one of the forward sections, and if anyone was going to get it first it would be us. As that thought ran through your mind, you didn't hesitate, you just went. As we swept round we were opened up on by a straggler. He couldn't have been a sniper or me and the section commander wouldn't be here today. He opened fire and we hit the deck. Soon afterwards we came under heavy mortar fire again. We tried to get some cover from the rocks, but even the best cover you could find would leave half your body exposed. I must have moved position four or five times. I just didn't feel safe anywhere. You just lay there wondering when yours was coming. That's a sound I'll never, ever forget – that heavy whistle of the mortar rounds coming in. If it sounded as if it were dropping short you wondered if it was yours. If it went over the top of you, you knew you were safe. As we lay there it never seemed to stop, it seemed an eternity, but can't have been more than about 25 minutes."

"When they finally stopped, and it all became stable, you just felt the exhaustion in yourself, you turned round and looked at each other and thought, 'My God, I'm still here!'"

"We were later able to rifle some of the Argie positions, some of their sleeping bags and blankets, just to sustain us. It was the first indication we'd had of how well they had been fed and supplied. We found supplies they'd left behind, that hadn't even been opened yet – food and ammunition – so all these stories that they were poorly fed and under-equipped were just not true, certainly not in the troops we came across."

Marine Dick Palmer, 'Y' Company 45 Cdo.

"During the attack on Two Sisters I was at the R.A.P. (Regimental Aid Post). I've always wondered how I would stand up to seeing a serious casualty and if I could apply the necessary first aid. I found out that all those boring lectures we'd had on the subject really did some good. When the casualties started to flow I helped get them all out of the choppers and helped where I could in changing dressings and so on.

After it was all over we had to put the dead into the choppers, breaking their bones to fit them all in. Then the Argie wounded came in, but now I had no sympathy for them at all after seeing the mess they had made of everything. I now know the cure for all battle casualties – confidence, ability and knowledge – or two aspirins sellotaped to the forehead! 'Non illegitimi carborundum'!"

Marine Shaw, 'Y' Company 45 Cdo.

"There was moaning coming from my front, about 10 yards ahead, so I crawled up to where it was coming from to find the boss, Chris and the gunner representative lying next to each other. They'd obviously been shot or hit by mortar rounds. Chris got up after I'd shouted at him, but the gunner and the boss were still. I checked the boss for injury and found some blood coming from his neck. He refused to move when I shouted at him, so he was obviously in pain but most of all very badly shocked. I got Chris to help me in dragging him back and then we went back for the gunner who was being lifted by one of his mates. Between the three of us, we managed to get him into cover behind a bank."

"By this time first aid was being rendered by L/Cpl. Forsyth on the boss while I attempted to revive the gunner, who had stopped breathing. Mouth to mouth and cardiac massage were given for about 30 minutes, to no avail, and then we covered his body with a poncho and marked his position."

"It was only in those few moments of calm later on as we were advancing that the raw truths of war were beginning to hit me."

Marine Ingles, 'Y' Company 45 Cdo.

A Marine for 21 years, 37-year-old Sergeant-Major George Meachin served in the Falklands with 'Yankee' Company 45 Commando. 'Y' Company 'yomped' from San Carlos to Two Sisters, where they fought their major set-piece battle of the campaign. George Meachin's pride in his men is heartfelt.

"They are 19-year-old average young men. They went down there as young – bumbling in the main – novice Marines who had barely joined the Corps. They had been nowhere, not seen much, half of them had seen Northern Ireland and Norway in the past, the other half had seen nothing."

"Running the company in the Falklands, looking after the men, was frustrating. Ammunition was a thorn in the side because of the re-supply chain – it was there, it came eventually, but the channel was blocked with obstacles. It was always worrying in case it didn't come. Then there was feeding the men. We were using Arctic rations which we were very much used to, but that was another concern. Then there was the well-being, and the equipment."

"'Casevac' (Casualty Evacuation) – a Sergeant-Major's responsibility at company level) was another problem. As always with a great force on the move, there had to be a slick, clean whisking away system for casualties. The word was that they stayed where they were. You left them intact on the side of the road. If a man fell by the wayside, from blisters say, or a bad leg or a wound, he stayed there. You tended him, you didn't abandon him, but he remained there to be scooped up by the medical teams."

"The Argentines! Were we guilty of underestimating them? Definitely! I came back from the Falklands, went to my local shop to buy some tobacco – full price, not duty free, and that was a shock I can tell you – and I was harangued by this little old lady. She was talking about us killing the little Argentine fellows, dark-featured, lovely fellows, who were mistreated, lacked food, were badly led, badly moraled. They were not badly fed. They were not badly moraled. They were not badly ammunitioned, they were not all conscripts. They were not all babies."

"When we attacked Two Sisters we came under lots of effective fire from 0.5 in. cal. machine guns, which claimed the life of one of my guys, 'Blue' Nowak. At the same time, mortars were coming down all over us, but the main threat was from these machine gunners who could see us in the open because of the moonlight. There were three machine guns, and we brought down constant and effective salvoes of our own artillery fire onto them directly, 15 rounds at a time. There would be a pause, and then they would come back at us again. So we had to do it a second time, all over their positions. There'd be a pause, then boom, boom, boom, they'd come back at us again. Conscripts don't do this, babies don't do this, men who are badly led and of low morale don't do this. They were good, steadfast troops. I rate them. Not all of them, but some of them."

Sergeant-Major George Meachin 'Y' Company 45 Cdo.

FROM THE SCATTERED MEMORY OF AN ARGENTINE CONSCRIPT

"For a long, cold and lonely week now we watched the British artillery strafe Mount Kent and Mount Longdon where 'B' Company was stationed. It was relentless, from sea, land and air. From our position, using binoculars, we could watch the movements of the British helicopters. They moved positions in the helicopters themselves, ferrying troops from one place to the next. On 10th June, we received orders to leave our positions, which were really quite comfortable, because the bombardments weren't punishing that area too much. We were ordered to rejoin the commando platoon where we'd started off, because they needed people to build new defences and to move the field guns. Most of our guns were facing east and now they had to be turned round. When we arrived at the commando platoon position, we saw the Harriers streaking low overhead. They gave them everything they had, they even used their Fal. Rifles. But the planes flew in over the channel, finding the exact position where shots from us, up on the hillside, would have hit our own troops below, and vice versa, those below could have hit us. The British ground artillery began pounding in the most incredible way, their aim was very accurate. Our 105 mm. guns, which had arrived twenty days earlier, responded quite effectively. We also received anti-aircraft Blow-Pipe missiles which are fired from the shoulder. The missiles track the heat of the plane's engine, but you had to know how to handle the electronic control. I saw N.C.O.'s who, with all the good will in the world, couldn't use them, and when they fired, the missile shot off in any direction, sometimes crashing into the ground. You can't start learning in the middle of a war."

"It was hell, bombs were falling on all sides of us, but we were lucky, we weren't hit. An N.C.O. was in charge of the gun, but an ordinary soldier lined it up, an absolute phenomenon, he shot very accurately. Over the radio, from a hill on the front line, the orders came: 'Shoot again, number three gun, shoot again, hitting the area well…' But then the British realised that the gun was causing them trouble, they began to hunt it down. This seemed to be their speciality, pinpointing and then hunting down. Where our positions were concerned the British Marines were in a class of their own as hunters. They had equipment which detected any kind of gunfire, even rifle shots, marking the exact co-ordinates of its position. Their artillery then began to aim for the trouble spot. The only possibility we had left was to change the gun's position, but it was impossible, it was very heavy. The bombs became more and more frequent, about one every two seconds. They were destroying all the mortars they detected. It was amazing. We must have fired twenty rounds at them with the field gun and they fired one hundred at us."

"The only thing we heard was the shouts of 'Watch out! Watch out!' as the bombs fell again. Some of them were dropping five or six yards from me; shrapnel was flying above my head. Even the smallest fragments were red hot. I was lucky, none of them hit me, but I saw them fall on the quilted anoraks of some boys next to me, and they just burned through everything, anorak, pullover, vest, right through to the flesh."

"The moment came when we just couldn't remain there any longer. The order to retreat didn't come, there were no officers to be seen near our position, so we had to take a decision. We were a group of about ten who had got stranded. We decided to retreat 500 yards in towards the hillside as it began to snow. It was the right decision; a few minutes later, we saw the cave we'd been in a few minutes before completely destroyed by the British artillery. The chunks of rock flew through the air. We were still cut off, even in our new position. We didn't know where the rest were, the groups had split up, people ran, wandered and got lost......"

"I was next to one of the walls and I was sort of plastered against it; another boy at the other end of the foxhole was sitting on a log and was hurtled through the air to land on top of me. The rocking was tremendous; it was like an earthquake, as if everything was going to fall apart. I thought my eardrums were going to burst. We'd been told that, in the case of a bombardment, we should open our mouths wide and try to scream because otherwise we ran the risk of going deaf. It was our first bombardment and we all tried to do what we'd been taught in a split second. Some shouted because we'd been taught to; others screamed from fear. I suddenly had terrible earache, it was as if liquid was pouring out of my ears; I tried to feel my face but there seemed to be nothing there."

Anon.

The above was related by an Argentine conscript after the surrender. His name and actual rank seems not to have been recorded. I tried but failed to find it.

"The Cost of Battle"

Within a group as small and homogenous as a ship, battalion or commando it is possible for the commanding officer to know something about each man. In 650 or 800 men there are the characters, the problems and the men who can be relied on to keep going in conditions of great physical and personal pressure.

Since the British forces are volunteers there is a stronger sense of cohesion than in a conscript army. Against this background the loss of men killed in action is profoundly felt. The Lieutenant, Captain or Colonel would know a great deal about the solder or sailor under his command. But, the hardest experience during the campaign fell to the mothers, wives and friends who were at home watching the news and listening to the, at times, ambiguous statements from the M.O.D. We learned this during the Falklands War but right here and now at the finish of Gulf War II (at time of writing) many facts, figures and statements of authenticity on whether or not we should have gone to war has got the public, in general, rather confused. Seven out of ten people, even those within the services, could not give even the mildest briefing as to what the Hutton Investigation is all about in any detail. There is this ongoing problem of trust and even many years ago Robert Fox wrote in 'Eyewitness Falklands' of a conversation with the wife of an officer in the Parachute Regiment after he had returned from the Falklands: "You bloody journalists out there don't understand what hell we went through back here, we wives. Can you honestly know what it was like looking after all the wives? To tell one with two babies crawling round her, and another on the way, that her husband isn't coming back!"

Telling the next-of-kin of the death of their son or husband fell to the men who remained at home. The Married Families Liaison Officer with the Royal Marines, and the Families Officer with an Infantry Battalion would take some of the weight of grief and worry. The families in the married quarters too were a source of strength – all of them sharing the fears and concerns with one another.

But for the young unmarried servicemen with parents living away from the barracks, home port or depot, the responsibility for breaking the news of a son's death would fall to the local police force. At the South Concourse of the Ministry of Defence the first public news of casualties came in typewritten and photocopied lists. They carried the full rank, Christian name, surname, age and home town of each man if he came from a ship, or arm of service if he was a soldier or Royal Marine.

All these experiences were in strong contrast to those of the Argentine conscripted servicemen who were flown to the Malvinas. When British-supervised burial parties started to collect the dead they discovered that some of the Argentine soldiers' identity tags only bore their blood group. The British forces carry two – each with the blood group, name, service number and religion. In the event of death, one tag is left with the body and the other collected as a record. When subsequent re-burial takes place the body can be identified. For many Argentine next-of-kin the fate of their husbands, sons and other relatives is still unknown, simply because the British undertakers who re-buried them at Darwin had no idea who they were. And so the story is told.

Standing among the shell-holes and shambles of battle, and watching the shocked, saddened faces of those who had lost their friends, one could echo the words of the 'Iron Duke' of Wellington who said: "There is nothing half so melancholy as a battle won... unless it be a battle lost."

2004 - THE FALKLANDS

Recently it was Julian Thompson who summarised things perfectly for the "Globe and Laurel", the Royal Marines magazine, when he wrote: "At 06.15, Falklands time, on 2 April 1982 Lieutenant Commander Giachino and four members of the Argentine Navy's Amphibious Commando Company walked up to the back door of Government House in Stanley with the intention of entering and demanding the Governor's surrender. The response was a burst of fire from the Royal Marines inside the house. Giachino and Lieutenant Quiriga were badly wounded, the other three hid but were later taken prisoner by the defenders. Giachino died of his wounds, the first death in the Falklands War of 1982. Within three hours the

Argentine 2nd Marine Battalion which had landed in AMTRACs at Yorke Bay on the Airport Peninsula had secured Stanley, and the Governor, Rex Hunt, had ordered the Royal Marines garrison to surrender to avoid further bloodshed.

We saw how all of this unfolded in our earlier chapters and as this progressed, in Plymouth they had been getting the show on the road for some seven hours before the first clash at Government House. At 03.15 hours BST (five hours ahead of Falklands time), Major General Jeremy Moore, then M.G.R.M. Commando Forces, had telephoned to say the Islands were about to be invaded and to bring the 3rd Commando Brigade to short notice, load and sail south.

Seven weeks later the landings at San Carlos took place. So began the land campaign in which the Royal Marines were to play such a key role. Well before this the S.B.S. had been carrying out their vital task of reconnaissance of beaches and landing zones, and of course they continued to operate until the end of the war. Without the Royal Marine crews of the eight L.C.U.'s and eight L.C.V.P.'s embarked in the two L.P.D.'s Fearless and Intrepid, no landings would have been possible. But this was only the beginning, these crews and those in the Raiding Squadron were indispensable to the success of the operation until the very last day of the fighting; their bravery and devotion to duty were an example to us all. While H.M.S. Antelope blazed in San Carlos Water, Colour Sergeant Johnston in his L.C.U. was alongside taking off her crew. Over the radio from Fearless came an order to bear off, as Antelope was about to blow up; but Colour Sergeant Johnston remained and took off the last survivors. Some two weeks later, Colour Sergeant Johnston, realising that urgently needed communications vehicles must be delivered to the 5th Brigade at Fitzroy, ignored instructions not to move in daylight, and was attacked in Choiseul Sound by four enemy Skyhawks. He and his crew were all killed. Colour Sergeant Johnston was awarded the Queen's Gallantry Medal posthumously. His was not the only recognition of the landing craft crews' gallantry, for Colour Sergeant Francis was awarded the D.S.M.

The Commando Brigade Air Squadron suffered casualties on the first day of the landings. Thereafter, despite the high risk from Argentine fighters, and especially Pucaras, the Squadron flew in all weathers, by day and night, to support the Brigade. Lieutenant Richard Nunn, flying forward to casevac the dying Lieutenant Colonel 'H' Jones from Darwin, was 'bounced' by a pair of Pucaras, shot down, and killed. His air gunner, Sergeant Belcher, survived, but lost a leg. Lieutenant Nunn was awarded the Distinguished Flying Cross posthumously.

Day and night, the Commando Logistic Regiment Royal Marines toiled to support the Brigade. Just before dusk on 27 May, Skyhawks bombed the B.M.S. Five men were killed and 27 wounded by exploding ammunition and missiles, including the netted loads of ammunition on the helipad waiting to be lifted to support the Goose Green battle. Despite explosions and flying shrapnel, which lasted through the night, the work of the Regiment went on, as it did in the Field Dressing Station where two unexploded bombs had lodged.

Royal Marines musicians in the medical support role served down south, in S.S. Canberra, and the hospital ship Uganda. No one who saw what they did could speak too highly of their professionalism. When time allowed, their music provided a welcome relaxation for frayed nerves and exhaustion.

To begin with the three Commandos were largely spectators of the Battle of San Carlos Water, fought between the Royal Navy and the enemy air force. At last the time to move out arrived, and 45 Commando began their great 'yomp' to Mount Kent, while 42 Commando in a series of sometimes hair-raising helicopter moves were lifted forward. "K" Company arrived in the middle of a fire-fight between "D" Squadron 22 S.A.S. and enemy special forces. 40 Commando were ordered to remain to secure the beach-head by Divisional Headquarters, who by now had arrived. The threat here was posed by either an enemy airborne assault, or Special Forces at Port Howard, who had been tasked to attack the back door while we were moving towards Stanley.

The two most formidable objectives forming the defences west of Stanley, Two Sisters and Mount Harriet, were allocated to the two Commando Units. A parachute soldier who accompanied Julian Thompson back to the Islands in 2001, took one look at them and said, "Thank God we were not invited to attack those." Both features, whose whale-backed ridges well over 1,000 metres long, were crowned with great dinosaur-like spines of rock, were strongly held. Thanks to excellent plans by both Commanding Officers, based on information brought back by the numerous fighting and reconnaissance patrols, the attacks were brilliantly successful. A Second World War general once said:

"The battles worth study and worthy of battle honours are not the bloody ones; they are the ones that yield victory with few casualties. It is the approach that determines the outcome, and we believed that our approach, our designs and our training, the skill of our individual (junior) Commanders in this most telling of fighting, this infiltration battle, would suffice for the difficult and unusual task that lay ahead."

He might have been speaking about Two Sisters and Mount Harriet. The courage, skill and motivation of every young Marine was tested again and again through the long hours of battle on those dark mountain sides; and on the subsequent days under enemy artillery fire, especially the 155 mm guns whose range of 24,000 metres allowed them to hit all over our hard-won objectives. The Argentines surrendered as the Brigade was readying itself for the final battle for Stanley.

Looking back from the vantage point of more than twenty years, it is clear that without the Royal Marines and the amphibious expertise across a wide range of skills, coupled with the two L.P.D.'s and their landing craft, no landing would have been possible; no repossession would have been achieved, and the Falkland Islands would be Argentine today. Britain would be a different country.

To complement what Julian Thompson has said, Major General Jeremy Moore continued this appraisal by reminding all of us as we celebrate the successful conclusion of a campaign during which the Corps, in just about every aspect of its work, had been at the heart of operations. Together with the Royal Navy and alongside the other services a victory was won which was to have much wider implications than were immediately perceived, and which went a long way beyond the objectives which the Task Force were given by the Commander, Admiral Fieldhouse.

First, it was immediately apparent that our country was not prepared to stand by and watch while territory, and more importantly people, for which and for whom we were responsible, were invaded and occupied against their will. This had many unforeseen benefits, though for anyone who studies history some, perhaps most, of them seemed to have been predictable.

One of the beneficiaries of Britain's actions was, ironically, the Argentinians themselves. After the war, General Moore received a number of letters from Argentina, and all of them were friendly, even grateful. In particular there was a card, which he received at Christmas from four young undergraduates at Buenos Aires University, who had served as conscripts in the war.

The burden of their message was that they wished to thank the members of the Task Force for demonstrating to them that their "Emperor had no clothes" – (the General's phrase) - and thus given them the encouragement to get rid of him; to insist upon the replacement of the Junta by a democratically elected Government.

Another group who have prospered in the aftermath of Operation Corporate are, of course, the Falkland Islanders themselves, as they are the first to acknowledge. Their economy has improved out of all recognition, at least in part because of the attention which the campaign brought to the region, and the development of tourism and fishing.

A third outcome was the effect which our rapid and robust action had upon the Soviet Bloc. Many factors contributed to the realisation behind the Iron Curtain that the West 'meant business', and in the end to Glasnost and the collapse of the system, but British action in 1982 was certainly one.

But it is the last matter which the General covered for the "Globe and Laurel" that is of most interest to all of us in the Royal Marines. Before they were repatriated to Argentina after the surrender in the Islands, many of the captured officers remarked that what gave the Junta the confidence to invade the Falklands was the conviction that Britain would not respond if they did. They drew this conclusion from the announcement by our then Secretary of State for Defence, that Endurance our Arctic Patrol Ship, was to be decommissioned, Fearless and Intrepid the Assault Ships scrapped, and the two carriers which actually fought the war sold or scrapped. They may have made a serious miscalculation, but had the invasion been mounted a few months later than it in fact was, we would indeed not have been in a position to respond. What is of significance today, of course, is that the services, and in particular the Naval Service, is specifically organised to provide a capability to project power wherever in the world our Government finds it in the national interest to do so. That is a maritime strategy.

In the case of the Corps, not only is 3 Commando Brigade a core part of this strategy, with the Commandant General as one of the Navy's two operational commanders; H.M.S. Ocean is in service, Bulwark and Albion, the replacements for Fearless and Intrepid, have been launched and are now commissioned, and the support vessels such as L.S.L.'s and auxiliaries are programmed.

More than twenty years after a defence policy which looked likely to spell the end for any amphibious operational capability for Britain, and a serious emasculation of the Royal Navy, the Senior Service looks better balanced and more capable than it has for a long time. All the branches of the Corps are looking well set to carry forward the traditions, the expertise and the fighting reputation of the Royal Marines into the future. An ironic outcome, one might think, to a stated policy which might well have spelt the end of our classic role.

OPERATION TELIC - THE LIBERATION OF IRAQ
Preface by The Commandant General Royal Marines
Major General A. A. Milton C.B., O.B.E.

The whole of the Royal Marines family will have watched events in Iraq unfold with great pride. The vast majority of the Corps, including significant elements from the Band Service and the RMR, has been involved in the past few months. The planning and preparations were conducted quickly but with great care and professionalism. The deployments after Christmas, 40 Commando Group by sea, the remainder by air, were defining moments. Even as the Brigade, and those afloat and elsewhere, went through an intensive period of preparations, often in very demanding conditions, it was still uncertain as to whether it would come to war. This was a very testing period for all; not least those back at home. In the event the Brigade were first into action, seizing the Al Faw peninsular in a highly successful landing against some difficult opposition. Subsequent actions were even more demanding and throughout the area the Royal Marines were carrying out vital roles. In the latter stages of the campaign I deployed to Bahrain as the Maritime Component Commander. I had the privilege of seeing Royal Marines with the Fleet and ashore with the Brigade carrying out their duties. I did so with great pride. I was particularly struck by the calm, measured professionalism and the restrained, self-effacing manner in which they described what in many cases had been fierce combat. Indeed I believe the press reporting, although very fair about the Corps, sometimes failed to convey just how intense some actions had been. I was also particularly impressed by the fulsome praise I heard from senior officers. As I write this in the Gulf, the process of returning home is underway. I know our hearts go out to the families and friends of those who lost their lives and were injured. I also know the Corps family will stand by them in the years to come. With Afghanistan last year and now this campaign, not to mention fire fighting and Northern Ireland, it has been a most demanding period for the Corps as a whole and our families in particular. However, it has also been a period of courage, commitment and very high professionalism. The whole Royal Marines family, serving, retired, friends and relations, can take credit for the fact the Corps has never been in better shape.

BUCKINGHAM PALACE.

7 May, 2003

Dear Commandant General,

Now that things are settling down again, I just wanted you to know how much I admired the way the Royal Marine Commando Brigade performed in Iraq. It must have been quite frustrating to have to hang about in pretty uncomfortable conditions before it all started and my impression was that accommodation cannot have been all that brilliant once the party started.

I am not aware of the precise details of the Brigade's employment, but that it achieved all its objectives with such a very small number of casualties is a great tribute to the competence of all its members. It was also quite evident that they performed their post-conflict tasks with great skill and tact. They all have every reason to feel thoroughly proud of their achievements.

Yours sincerely

Philip

Foreword by The Commander 3 Commando Brigade Royal Marines - Brigadier J. B. Dutton A.D.C.

Planning for what eventually became Operation Telic started in September 2002 and for three months was based on a single Commando (first 45 Commando, later 40 Commando) and an Amphibious Ready Group. Changing enemy force levels, together with a US request for us to take on greater responsibility, led to the commitment in early December 2002 of 3 Commando Bde RM(-). 45 Commando RM was not involved; they were committed to the operation, but not to the Brigade. In the event, W and X Companies played a vital role elsewhere on the operation, Z Company did reinforce the Brigade while Y Company held the fort on Operation Fresco. To balance the Brigade, the USMC placed 15 MEU (SOC) under command for the initial phase of the operation. The Brigade deployed by sea and air; 40 Commando Group in amphibious shipping on 16 January 2003 and the remainder, including 42 Commando, by air direct to Kuwait between 20 January and 15 February 2003. Initially the Brigade worked direct to 1 MEF, but when the UK national main effort shifted from the northern approach through Turkey to the southern approach through Kuwait in early January, HQ 1 (UK) Armoured Division was interposed. We then worked alongside 7 Armoured Brigade and 16 Air Assault Brigade.

The initial operation was a complex, opposed helicopter assault at night onto the Al Faw Peninsula to seize the oil infrastructure intact and to provide flank protection of the MCM force clearing the Khawr Abd Allah waterway to Iraq's only deep water port, Umm Qasr (UQ). As the first ground force action of the war it had strategic significance. It spanned all four environment components – Special Forces, Maritime, Land and Air – without the benefit of an Amphibious Objective Area (AOA) to aid deconfliction. It involved integration with the USN Special Warfare Group, the SEALs. It required a considerable Offensive Support effort, including AC 130 gunships, US close air support, US and UK artillery and UK and Australian naval gunfire support.

Following this initial operation on the Al Faw (and in Umm Qasr by 15 MEU) the Brigade advanced and was involved in a series of engagements, culminating in the fall of Basra. Succeeding articles tell these stories graphically and I will not attempt to describe them all here. I will confine myself to a few observations. The loss of the US CH46 helicopter on insertion, which killed eight members of the Brigade and the rocket attack on the LCVP, which killed a Marine from 9 ASRM were tragic. Nevertheless, the operation overall was completely successful. Whilst the level of resistance put up by the enemy eventually proved to be quite low, with the threat being largely asymmetric, we did not know that this would be the case before battle was joined: the left and right of arc of potential enemy reaction were far apart. Certainly, on 20 March as we made final preparations in the tactical assembly area (Viking) to launch the assault, we fully expected to be subjected to chemical attack and for the helicopters to be engaged by air defence artillery. We had to assume a determined resistance and, at local level we met it, against some determined and fanatical fighters. Although the enemy did succeed in mounting several armoured attacks out of southeast Basra, his failure stemmed from his inability to put together a 'joined-up' all arms defence. I attribute this failure in part to the surprise engendered by the speed and violence of our assault. I am delighted that we won our battles convincingly and with minimum casualties.

Of course we are capturing the lessons identified from this operation, but there are a number of facts and achievements, which are worth noting:

• We conducted the first conventional ground force action of the war.

• We launched a two Commando Group aviation assault.

• We conducted an opposed landing, contrary to Defence Planning Assumptions.

• This was the first use of Naval Gunfire Support by the UK since the 1982 Falklands War and the first for Australia since the Vietnam War.

• The operation involved a Command and Control structure that spanned all four environmental components.

• 40 Cdo provided support to SF (in this case the US Navy SEALs).

• 15th MEU (SOC) USMC was placed under command of the Brigade.

• A Challenger 2 Squadron and an Armoured Recce Squadron were integrated into the Brigade.

• Our operation involved the first-ever operational use of an M3 ferry crossing.

• It again proved the sea-basing concept for aviation and logistics.

To expand on this last point, although this was not a classic full scale amphibious landing, it would not have been possible without the support of the ships. They provided a base for 40 Commando until 48 hours before the assault (and for one company and the majority of vehicles and stores, throughout). They also provided basing and maintenance facilities for 845 Squadron Sea Kings and 847 Squadron Lynx and Gazelles throughout. The value of this was evident in the much higher availability for these aircraft compared to those based ashore. Lastly, but very importantly, 1 (UK) Division could not have conducted this operation without the use of the sea-based stocks. The whole Division was sustained until after the conflict had begun by drawing on the afloat stocks designed for just 3 Commando Brigade.

As ever the overriding factor in the success of this operation was the quality and endurance of the marines, soldiers and sailors of the Brigade and of all our attached units, including the reservists who fitted into the team with ease. All our plans would have been valueless if they had not executed them so superbly – with bravery, enthusiasm and cheerfulness. There was a willingness to close with and kill the enemy. The foundation of commando training is willpower and teamwork, and so it proved here. In short, our training worked. Furthermore, marines were able to adjust from war fighting to peace keeping in a matter of hours. Command training from the JCC up to Staff College provides leaders who can plan and command operations across the spectrum of conflict. We have sometimes been criticised for 'gold plating' at CTCRM and I appreciate that collective training such as Tesex and Ex Black Horse with the USMC are expensive, but it is worth it and I will continue to defend these standards strongly – they saved lives.

OPERATION TELIC – DIARY OF KEY EVENTS

The following is a synopsis of operations conducted by 3 Cdo Bde RM during the first two weeks of the war, summarised from the reports emanating from various Headquarters and the media, in chronological order.

Preparations

On arrival in theatre, operational control of the Brigade was transferred from the Commander-in-Chief FLEET (CinCFLEET) to that of Commander 1 (UK) Armoured Division (1 (UK) Armd Div). As such, the Brigade Commander reported through the Land Component chain of command. In the early stages, 3Cdo Bde RM took two additional elements under command; the USMC's 15th Marine Expeditionary Unit (15 MEU) and a Squadron of CVR(T) light tanks from the Queens Dragoon Guards (QDG). Both were fully integrated into the Brigade through intensive training in the period leading up to hostilities.

Following an extensive training package in Kuwait, the Brigade moved forward to Assembly Areas in the final days before 20 March 2003. Despite having sailed into theatre with the Amphibious Task Group (ATG), 40 Cdo RM flew ashore in the days preceding the assault, leaving only D Coy aboard HMS Ark Royal to provide an immediate alternative should poor weather prevent flying from Kuwait. In addition, given an initial plan to offload the QDG by USMC LCAC (H) Hovercraft on the Al Faw Peninsular, the QDG remained at sea aboard the USS Rushmore during the initial stages.

In the immediate lead up to the assault, 7 and 8 Batteries of 29 Commando Regiment Royal Artillery (29 Cdo Regt RA) were joined by a Battery of AS90 Self-Propelled Guns to form an Offensive Support Group (OSG). That OSG moved from mainland Kuwait to Bubiyan Island on 18 March to establish gun lines in support of 3 Cdo Bde RM objectives in the southern part of the Al Faw Peninsular.

The contribution of the Commando Logistic Regiment (CLR) must also be noted. Although designed to support the Brigade exclusively, CLR sustained most UK Land Headquarters, Units and Formations through the early stages of the deployment until those elements became self-sustaining. More specifically, by 19 March, CLR had ensured that all Brigade Combat and Combat Supporting Units were fully equipped for operations.

Thursday 20-Friday 21 March 2003

At approximately 20.00 hrs, 40 Cdo RM landed by helicopter assault in the area of an oil pipeline south of Al Faw town with the strategic objective of securing those oil facilities before they could be destroyed by Iraqi Forces. As such, 40 Cdo RM were the first non-Special Forces troops to enter Iraq. The unit encountered light opposition from Iraqi irregular forces, which it defeated and despite some being prepared for demolition, all objectives were secured without loss or destruction. The significance of this success should not be underestimated given the ecological disaster that would have resulted from the demolition of those facilities.

In the early hours of 21 March, 42 Cdo RM landed, also by helicopter, on the Al Faw Peninsular. The unit landed to the north-west of 40 Cdo, establishing a 'blocking' position to prevent enemy interference with 40 Cdo's objectives. Simultaneously, 15 MEU crossed the Kuwait-Iraq border moving into the town of Umm Qasr, their objective being to secure the area to enable humanitarian aid to be landed in the port facility as quickly as possible.

The Brigade Reconnaissance Force (BRF) also landed by aviation assault on the Al Faw Peninsular alongside 42 Cdo RM. During the course of that move, a USMC CH-46 Sea Knight helicopter crashed just short of the Kuwait-Iraq border, killing all on board. The cause of that tragedy has yet to be established although there is no suggestion it was attributable to enemy action. It was with the utmost regret that the Brigade reported the loss of Maj Jason Ward, Capt Phil Guy, WO2 Mark Stratford, CSgt John Cecil and Marine Sholto Hedenskog from UKLF CSG and Sgt. Les Hehir, L/Bdr. Llewelyn Evans and OM(C)1 Ian Seymour from 29 Cdo Regt. RA in that accident.

Throughout the assault, indirect fire support was provided by the OSG located on Bubiyan Island together with a four-warship gunline comprising HMS's Chatham, Richmond, Marlborough and HMAS Anzac. Their fire was co-ordinated by officers of 29 Cdo Regt. RA.

Saturday 22 March

40 Cdo RM continued the clearance of the southern tip of the Peninsular, including the town of Al Faw. The unit continued to meet and defeat light opposition throughout. Of particular note, A Coy

conducted a night assault against a defended building in Al Faw town during which two Marines received burns. Although serious, neither injury proved life-threatening.

Simultaneously, from its 'blocking' position, 42 Cdo began moving west, clearing the northern bank of the Khawz-Al-Arab (KAA) waterway aiming to link up with the CVR(T) of the QDG who had by now crossed the border on land. The aim was to clear the bank of enemy to enable mine-clearing of the KAA, in turn to enable the movement of humanitarian aid into Umm Qasr. Throughout this period, 15 MEU continued to quell pockets of Iraqi resistance in both the town and port areas of Umm Qasr.

Following the initial landings, the additional Battery of AS90's returned to the command of 1 (UK) Armd Div to contribute to the advance of 7 Armoured and 16 Air Assault Brigades. The Batteries of 29 Cdo Regt RA moved from Bubiyan Island to the mainland and then by land to the Al Faw Peninsular.

Sunday 23 March

Having completed the clearance of Al Faw town, 40 Cdo moved north-west, clearing the southern bank of the Shatt Al Arab (SAA) waterway which forms the international border between Iraq and Iran. Reaching their established limits in the north of the Brigade Area of Operations (AO), 40 Cdo RM joined the BRF (now incorporating the QDG) in establishing a 'blocking' position across access routes to the Al Faw Peninsular to prevent enemy incursion from the city of Basra. That position was established by last light. Meanwhile, 42 Cdo continued to occupy the northern bank of the KAA, providing security for the Royal Navy's mine-hunters at work in the waterway.

Monday 24 March

An Iraqi Armoured Battalion, comprising some 50 T-55 Main Battle Tanks (MBT's) advanced south from Basra into 3 Cdo Bde RM's AO. 40 Cdo RM and the BRF, now supported by 29 Cdo Regt RA, tasked a variety of air and aviation assets in joint attacks against that force. The action resulted in over 20 enemy MBT's being destroyed with the remainder withdrawing into Basra.

Simultaneously, at 02.00 hours, 15 MEU commenced a move north along the Shatt-Al-Basra (SAB) to secure the town of Kaz, the final 3 Cdo Bde RM objective for the initial assault. That objective was secured on 25 March. At the same time boats of 539 Assault Squadron Royal Marines (ASRM), with support from 42 Cdo RM, began clearing the waterways north of Umm Qasr.

Tuesday 25 March

Having commanded operations from Northern Kuwait throughout the initial stages, the Brigade Commander moved his Headquarters into Southern Iraq on 25 March. 40 Cdo RM and the BRF continued to maintain their 'Blocking' position in the north to prevent incursions of Iraqi forces onto the Al Faw Peninsular.

Having cleared the town of Kaz, 15 MEU were detached from 3 Cdo Bde RM to join USMC forces conducting operations further north. They were relieved in Umm Qasr by 42 Cdo RM, now freed from protecting Mine-Counter Measures (MCM) operations in the KAA. 42 Cdo RM immediately commenced a highly aggressive search and patrol routine across Umm Qasr that provoked a number of minor contacts with the enemy that were prosecuted without loss.

Wednesday 26 March

On 26 March, 14 Challenger II MBT's from 'C' Squadron of the Royal Scots Dragoon Guards (RSDG) were attached to 40 Cdo RM to assist with maintaining the security of the Al Faw Peninsular. Together with the BRF, 40 Cdo RM continued to maintain the blocking position, effectively preventing enemy forces from moving south from Basra, thereby 'fixing' the enemy.

42 Cdo RM continued to aggressively patrol Umm Qasr, maintaining the security of both port and town to enable the safe docking of the RFA Sir Galahad, already laden with humanitarian aid for the region. The arrival of Sir Galahad was delayed by some 24 hours with the finding of further Iraqi mines in the waterway.

Thursday 27 March

In the north of the Brigade area, the enemy launched another armoured attack from Basra, comprising 14 T-55's, against 40 Cdo's position. On this occasion, the Challenger 2's of C Squadron RSDG, under the command of CO 40 Cdo RM, destroyed all 14 enemy tanks without loss.

South of 40 Cdo RM, 42 Cdo conducted a number of complex tasks:

• The unit provided support to 539 ASRM in a pre-planned clearance of known enemy locations in the area of Umm Khayyal.

• Having identified a position of approximately 30 enemy some 10 kms north of Umm Qasr, 42 Cdo RM launched a Company attack against it. That assault was highly successful and conducted without loss.

• The Unit also began the process of humanitarian support by delivering water to villages outside Umm Qasr.

• Within Umm Qasr, '42' also supported the Deputy Brigade Commander as he began to co-ordinate the delivery of humanitarian aid and to bring a semblance of normality back to the town.

Friday 28 March

28 March was a relatively quiet day across the Brigade AO, the highlight of which was the arrival of the RFA Sir Galahad in Umm Qasr carrying humanitarian aid. This enabled the Deputy Brigade Commander to begin providing assistance to the local population.

Saturday 29 March

The offload of humanitarian aid from RFA Sir Galahad in Umm Qasr was completed. In addition, aid convoys from Kuwait were escorted into the towns of Umm Qasr, Al Zubayr and Safwan.

At home, the bodies of those UKLF CSG and 148 Battery personnel lost with the CH-46 accident on the night of 20/21 March were returned to the UK at RAF Brize Norton. An emotional and very sad occasion; the event was attended by the families of those killed. The Corps was represented by CGRM, by Pallbearers from Parent Units and a RM Band.

Sunday 30 March

In the early hours, 40 Cdo, supported by L Coy from 42 Cdo, commenced an assault on an Iraqi Battalion to its north in the town of Abu Al Khasib, some 10 kms south-east of Basra. Codenamed

Operation James, the aim was to close access routes to the Al Faw Peninsular while further degrading Iraqi forces. A 15-hour battle ensued during which 40 Cdo took over 200 enemy prisoners of war including a number of Iraqi senior officers. Although sustaining a limited number of injuries from vehicle accidents and indirect fire, 40 Cdo RM suffered no deaths. The battle has been widely heralded in the national media.

Simultaneously, the remainder of 42 Cdo continued to patrol Umm Qasr and the waterways of the interior, hunting down Surface to Surface Missile (SSM) launch sites. That process was enabled by the boats of 539 ASRM and 9 ASRM (from HMS Ocean) and assisted by Z Coy, 45 Cdo RM. During that clearance, a Landing Craft (LCVP 5) from 9 ASRM was hit by a rocket propelled grenade that caused five casualties. Regrettably, Marine Chris Maddison died of his wounds.

It was with great regret that the loss of Maj. Steve Ballard of the UK Landing Force Command Support Group (UKLF CSG) was reported on 30 March. Maj. Ballard collapsed suddenly and died whilst deployed in theatre.

Monday 31 March

In comparison to recent events, this was a relatively quiet day across the Brigade AO although 2 SSM's impacted within the Brigade area. 40 Cdo RM concluded Operation James, securing their gains, regrouping and administering the companies. 42 Cdo RM continued to patrol Umm Qasr with great determination, finding a cache of 23 Iraqi sea mines in containers. 539 ASRM, supported by Z Coy, 45 Cdo RM, continued to patrol and secure the waterways to the north of Umm Qasr finding a number of small boats, some of which contained weapons.

Tuesday 1 April

A relatively quiet 24 hours in theatre in which 40 Cdo secured their gains of recent days in the Abu Al Khasib area, inevitably taking the opportunity to draw breath while the campaign progressed in other directions. 42 Cdo continued to patrol both Umm Qasr and the waterways and marshes to the north with the specific intent of finding the launch sites of Surface to Surface Missiles (SSM's) that have been threatening both Coalition Forces and neighbouring nations. In Umm Qasr, the Deputy Brigade Commander and a small staff continued to co-ordinate the distribution of humanitarian aid to the local population. In the meantime, 29 Cdo Regt. RA engaged a number of enemy mortar positions with Counter-Battery fire overnight.

In less exposed areas, units now patrolled in berets, as opposed to helmets, to defuse tension. It was hoped that this would encourage the local population to interact more readily and thereby accelerate the return of the region to some semblance of normality.

Wednesday 2 April

The Brigade commenced a rolling 'refurbishment' programme. In appropriate numbers, units withdrew personnel to the Cdo Log Regt RM where weapons, vehicles and equipment were serviced and individuals were able to receive centralised feeding and a period of uninterrupted rest. This did not interrupt the Brigade's ability to prosecute offensive operations at a high tempo.

40 Cdo remained 'firm' in securing their most recent gains. 42 Cdo continued to patrol aggressively to isolate and destroy any remnants of resistance. The Unit raided a specific objective capturing 7 Ba'ath Party activists in the process. At the northern end of Umm Qasr, a crowd of some 400 locals, seeking to loot humanitarian aid, was dispersed by Royal Marines. Measures were taken to prevent such concerns in future.

7 and 8 Batteries of 29 Cdo Regt RA remained alert in providing surveillance assets and indirect fire support as necessary. 539 ASRM moved a significant number of boats overland to the Shatt Al Arab waterway, which runs into the centre of Basra. Together with Z Coy, 45 Cdo RM, '539' patrolled waterways across the full breadth of the Brigade area of operations.

Thursday 3 April

C Sqn QDG were rouled through Cdo Log Regt over 24 hours, to service equipment and weapons, to rest and for centralised feeding. The Brigade Commander took the opportunity to issue every man with the 3 Cdo Bde RM 'flash' in recognition of the Squadron's outstanding support over the previous 14 days.

40 Cdo RM remained 'firm' in the north of the Brigade area of operations. 42 Cdo RM continued to patrol robustly and successfully prosecuted a small number of contacts with irregular enemy forces. Having extended their operating base, 539 ASRM continued to dominate waterways within the Brigade area.

Notwithstanding an alert posture, the Brigade seized every opportunity to defuse tension. Though a delicate and difficult balance, Royal Marines began to establish links with the population. Deliberate measures included the wearing of berets rather than helmets, the distribution of humanitarian aid and the provision of engineering support and medical assistance to civic services. Royal Marines played two football matches against local teams. J Coy 42 Cdo RM were defeated 9-3 by the Az Zubahir team and in the other, the Corps lost again, this time 7-3, to the local Umm Khayyal side. In all probability the result of conversations between young Marines and locals, it again demonstrated the quality and sensitivity of our men in understanding their environment. Notwithstanding such dents to the Corps' sporting reputation, such events play a fundamental role in engendering trust. Inevitably, CO 42 Cdo demanded a re-match.

Friday 4 April

During the course of the previous week, surface temperatures rose significantly across the area of operations. At this stage, troops were operating in temperatures in excess of 38 degrees Centigrade.

From the Brigade HQ perspective, the past 24 hours were spent deconflicting and co-ordinating with the other formations within 1 (UI) Armd Div. Challenger 2s of the Falcon Sqn of the Royal Dragoon Guards (RDG) relieved those of C Sqn Scots Dragoon Guards (SCOTS DG), under the tactical command of 3 Cdo Bde RM.

All Brigade units were firm within respective areas, each conducting extensive patrol programmes to dominate the area, prevent activity by Ba'ath Party activists and making every effort to establish links with the locals to restore civic normality as quickly as possible. In that regard, every effort was being made to respect the sanctity of Moslem Holy sites and also in that vein, 539 ASRM established links with local Iraqi fishermen on the waterways during the course of riverine patrols.

C Coy, 40 Cdo RM spent 24 hours at the Cdo Log Regt RM's 'refurbishment' centre.

This was a highly successful period of operations by 3 Cdo Bde RM. The units of the Brigade achieved all that had been asked of them and a great deal more, and this was reflected in intense media coverage. The importance of the Brigade's intervention during the early stages, in preventing environmental disaster in the Northern Gulf, is of particular strategic importance. Subsequently, RM units conducted operations spanning the full spectrum of conflict from full-scale warfighting on the one hand to humanitarian operations on the other. This was achieved often by the same men, on the same day and in the face of a populace from which friend and foe are typically indistinguishable. It is a matter of the utmost pride that the Corps continues to train individual marines that are capable of such adaptability.

The Brigade settled into a more routine programme of patrolling and assisting in the distribution of aid, continuing to make its presence felt across its area of operations for very different reasons. There was a shift of emphasis from warfighting to the establishment of 'normality'. Units went far beyond that, building relationships with the local population to defuse tension and inspire trust. This is a skill in which our men excelled and the response was most encouraging. Nevertheless, it must be clearly understood that the Brigade was operating in a volatile and fluid environment in which the possibility of contact with an irregular enemy remained a constant threat. In addition, notwithstanding the Brigade's good work in inspiring confidence, there remained significant objectives to be achieved. There was no illusion as to the potential for further offensive operations nor as to the Brigade's immediate ability to adopt a more aggressive posture in order to prosecute them.

Despite the outstanding operational success, it has been a time of the utmost sadness for our losses and overwhelming concern for the families that must inevitably bear the greatest pain. Our thoughts and prayers go to those of the Corps family who have suffered throughout this conflict.

Diary compiled by Lt Col David King RM, SO1 O & D, Director Royal Marines.

"A Short Thinking Intermission"

Reading this book will not substitute in depth and intensity for the experience of serving in the Royal Marines, but there are powerful lessons here to take pride in and to guide and sustain us in every confrontation, whether at junior or senior level. In times that call for defiance and boldness there is no better role model for the warrior hidden in each of us than the one described throughout the actions and events of these pages. I, as a former Royal Marine, can only hope that every serviceman and woman will read – and reread – this book, place themselves in their imagination into every action described and actually live them, understand them, and take their message to heart. Remember my words of an earlier chapter "the trick is to stay alive".

The thoughts contained here represent not just guidance for actions in combat, but a way of thinking in general as we forge our way through the minefields of battlefield etiquette and legality in this wonderful new century where a policeman, dare I say, hardly has room to reprimand a thief without putting at risk his job, pension, mortgage and everything else he holds dear, if he is not completely up to date with the idiosyncrasies of the human rights 'card' players. In this book I have tried to bring to the surface and describe the Royal Marine philosophy for action which, in war and in peace, in the field and in the rear, dictates his approach to duty.

Long before I had reached the Kings Squad in basic training I had, through observation and common sense, learned the Corps' style of warfare required intelligent leaders with a penchant for boldness and initiative down to the lowest levels. Boldness is an essential moral trait in any leader, for it generates control beyond the physical means at hand. If you cast your mind back to our first Northern Ireland section (chapter 20) when we took possession of the Mount Norris bombers I, in fact, left myself wide open to monumental 'cock-up' as they say. The situation on the ground was wild and wide on terms of legality. Had the bombers made a collective decision to mount a challenge I firmly believe I would have served a long prison term, as there was no way I would allow these six players to walk. Had they succeeded with their plan (they had enough material for at least a 100 lb bomb) in 1974, who knows what supermarket or bus station would have been littered with the dead.

Would my action in shooting one of them in order to frighten the others into inaction be looked upon as boldness or self-destructive stupidity? We must remember these traits carried to excess can lead to rashness, but we must also realise that errors stemming from over-boldness are a necessary part of warfare. Not only must we not stifle boldness or initiative, we must continue to encourage both traits in spite of mistakes which are sometimes made in many theatres of operation.

On the other hand I am highly adamant on dealing severely with errors of inaction and timidity. We must never accept lack of orders as justification for inaction; it is each and every Royal Marine's duty (no matter what his rank) to take initiative as the situation demands. I did on that evening at Mount Norris and I called a successful bluff. But even today I can't help thinking how damned lucky I was not to have been forced into playing with that derogatory human rights card.

Trust is another essential trait among leaders – trust by seniors in the abilities of their subordinates, and by juniors in the competence and support of their seniors. Trust is a product of confidence and familiarity. Confidence among Marines of all grades results from demonstrated professional skill. Familiarity results from shared experience and a common professional outlook on life of men in uniform for the long term. Sometimes the very long term. Relations among all leaders – from Corporal to General – should be based on honesty and frankness, regardless of disparity between grades.

Much earlier in this book I mentioned meeting Field Marshal Lord Montgomery of Alamein in Germany in 1964. I was on the verge of leaving the Royal Warwickshire Regiment with which Monty had had a life-long association and he was on a farewell visit to the unit before his transfer from uniform to red-robe status in that daunting palace on the River Thames. He was a man renowned for not partaking in the flummery of long drawn out inspections unless absolutely necessary so, true to his form, he drew the Battalion around him in a huge informal 'gang'. He had been informed by the C.O. Lt. Col. H. Illing that two junior command courses were on parade and almost finished their training. As he spoke to us it was clear that most of his remarks were aimed at them.

The first thing that struck me about Monty was what a very slight man he was. I am only 5'10" and throughout his perambulations back and forth, on the occasions when he stood in front of me I found myself, quite frankly, looking down at what came across to me as a hugely calm and contented man who looked as though he wanted to say he was sorry to put us to all this trouble on a morning such as this. He had a gait coupled with some traits that he held even as he walked off the parade ground some half-hour later. His left forearm was out of sight permanently behind him resting in the small of his back and in his right hand he held the surprisingly short Field Marshal's baton in an upward position which placed the figurine of Saint George slaying the dragon embossed on its tip just under his right ear resting by the side of his cheek. He looked as though he had tried and tested this position many years ago and was completely content with it. An earned sort of contentment from knowing he would never again have anything to prove to anyone.

Monty decided to finish off this farewell parade with a few words which were aimed primarily at the up and coming junior NCO's when he said, "The purpose of all training is to develop forces that can win in any combat situation. Training is the key to all combat effectiveness and therefore is the focus of effort of any peacetime army. However, training should not stop with the commencement of war; training must continue during any new hostilities to modify to the lessons of combat. Professional military education is designed to develop creative, thinking leaders. A leader's career, from the initial stages of leadership training, should be seen as a continuous, progressive process of development. From the rank of junior NCO to each stage of his career, he should be preparing for the subsequent stage."

Monty went on for another ten minutes or so but despite the freezing temperature of around minus five, I think he got the message across that morning so long ago on that parade ground in Germany. For basic 'boots on the ground' foot soldiering today, his advice and thoughts still stand sound. So, as I close with this slight diversity for reminiscence let us move forward now – or should I say let us move backward to the skills of the real hunter class.

CHAPTER 26

PATROLLING AND TRACKING

"In warfare, patrolling is the basis of success. It not only gives eyes to the side that excels at it, and blinds its opponents, but through it the soldier learns to move confidently in the elements in which he works." This quotation from Field Marshal Viscount Slim of Yarralumla, KG, GCB, GCMG, GCVO, CBE, DSO, MC, KSTJ, is appropriate because it was made by an outstanding soldier, whose greatest victories were won in South East Asia. As it emphasises, the key to success in fighting on most terrain can depend on successful and experienced patrolling. The history of World War Two has shown that a unit that patrols efficiently will have mastered the skills and techniques that are necessary to ensure its success in any other theatre of war.

More contemporary history in Korea, Malaysia and in fighting communist revolutionary warfare in Vietnam has added to this lesson by stressing the need for not only the infantry, but for all arms and services, in both forward and rear areas, to be able to protect themselves by patrolling actively. In an operational environment a commander will constantly seek to maintain the initiative, to gain information about the enemy and to ensure the security of his own force. These requirements can be met by active, aggressive, and successful patrolling.

Within the Royal Marines new equipment and weapons continually improve a commander's ability to obtain intelligence of, and to inflict damage on, the enemy. Nevertheless no matter how modern or technical war becomes, the basic infantry patrol continues to increase in value because it is limited only by the ingenuity with which it is employed and the skill and aggressiveness of its members.

During my periods with 40 and 42 Commandos at Simbang in the 60's my training sessions at JWS (Jungle Warfare School) soon made me realise that the undulating terrain and dense vegetation frequently found in Malaya and most of South East Asia limit visual reconnaissance, surveillance and good observation. The concealment offered by such conditions is a boon to enemy infiltration, penetration, surprise attack and insurgent activity. A large number of foot patrols is needed to gain information and to provide security to defensive positions, harbours and bivouac sites. Many of these patrols we carried out in daylight. Where the terrain restricts movement by vehicles to a few roads or tracks there is every possibility a large number of foot patrols could be needed.

If careful thought and consideration is not given, enemy infiltration and insurgent activity will threaten installations and units in rear areas as well as forward units. During most training periods I went through at JWS, it was the Gurkha soldiers who acted as enemy and believe me, the elusive character of these professional "insurgents" required all units, irrespective of whenever or wherever we were located, to patrol actively by day and night for our own protection. Patrolling is carried out, as we have said, by both sides by day and night in almost all phases of war. It is often difficult and always dangerous.

It is easy for me to look back now and to fully understand what Monty was trying to tell us that cold morning in Germany. I thought then he was directing this advice at the two junior command courses, but I know now it was directed at every foot-soldier of infantry kind. Successful patrolling calls for a high standard of leadership, skill in weapons and fieldcraft, discipline, good team work, initiative and determination. It has a beneficial effect on our morale and will adversely affect the enemy's morale.

To put it quite simply the aims of patrolling are:

- To gain and retain the initiative
- To gain information
- To deny information to the enemy
- To harass the enemy
- To provide security
- To provide protection

Irrespective of their specific tasks, all patrols are a source of that most important of all elements, intelligence, and they must remember and report all information discovered.

Another point to ponder from my own personal experiences of patrolling through desert, jungle, arctic and urban European-type theatres of operation with the Royal Marines was how great an influence the climate always played. But the problems of operating in Iraq's seemingly endless summer heat was not unprecedented. My patrols in Aden in the 1960s for example, were planned to last ninety minutes, this being the maximum time it was thought men could stay fully alert at very high temperatures. But these things, more often than not, were very elastic in an active service environment. Things could happen to make patrols shorter or sometimes much longer. Remember we spoke earlier of "events, dear boy, events", and so the U.S. forces must have found sustaining alertness very difficult as they walked into trap after trap. This was obvious in the U.S. regions not from the lack of manpower, but from their lack of experience in dealing with the public who were in turmoil and the armed gangs, to put it more to the point, who left the U.S. troops with no uniformed or distinguishable enemy. Unlike the British in the south of the country, the Americans were to learn that coming to terms with the rules of engagement when facing warring civilians was an art that takes years of experience on the street "boots on the ground", and not locked in armoured personnel carriers. I believe in their early stages of securing the streets they came to realise they could either go the British way and observe the rules of engagement, or go their own preferred way of slipping their safety catches at the slightest shadow and hosing everything straight to hell in a handcart.

I recently asked some Royal Marines who shall remain anonymous, for their views on this and they answered me in all honesty by saying: They do it because they can. They can engage, time and again, in monumental cock-up, apologise and walk away unscathed. No-one will hold them to task on their actions.

They watched the British become embroiled in a God-awful situation with civilians in Derry on Bloody Sunday when 14 people were killed and they see us now almost thirty five years later ensconced in never-ending fruitless debate having spent some one hundred and fifty million pounds, and counting in costs so far. When the current enquiry finally reaches a conclusion, the parties whose interest it is in will then condemn it as a "whitewash" and demand, as is their right, a new enquiry and so the old story rolls on and on. This, of course, teaches the Americans, Israeli and many other nations a thing or two about trying to do the democratic thing. Engaging in political correctness at all cost while the rest of the world, by and large, falls around laughing at the shackles successive governments continually forge for the command structure of our armed forces, especially the infantry at the coal-face of field and urban operations.

This, as a Royal Marine, I always found to be a great hindrance wherever or whenever paramilitaries in Northern Ireland or "regime loyalists" as they now call them in other parts of the world move into the streets for ongoing operations, or more likely in today's political climates in many countries, "to fill an authority void". When it becomes self-controlled the Iraqi security forces will have a long and bloody internal struggle on their hands. Looking at the situation now in 2004 it literally throws me back in years to late 1969 in Northern Ireland when parts of Belfast were in smoking ruin and the R.U.C. lay defeated and in disarray on the streets just waiting for the military to stamp its authority. That, as now in Iraq, is the time for serious and above all experienced urban patrolling.

As we have yet to learn in Northern Ireland – yes, even after all these years of defeatist political correctness – security voids cannot be managed from a distance. The TV, radio and press in that province report, almost on a daily basis, horrors which never reach mainland Britain. It is considered we don't "need to know"! Both London and Dublin maintain the deal is done. The situation is resolved. Let's hear no more about it. We can manage the security void from a distance. I'm sorry, I keep coming back to Ireland, but I cannot help making comparisons to my first tour of duty in late '69 when we were told we would be there only until the R.U.C. rediscovered law and order. Well, that was thirty five years ago now and I believe........what do I believe? Just let me repeat those words of good old Ronald Regan who said: "You ain't seen nothin' yet." The third world is awakening.

I am making the above comparison to the violence of the sort endemic in today's Iraq where the visible presence of coalition forces is likely to be required for years to come, regardless of the private "security" companies people talk about. There are lethally strong and cunning tribal candidates out there awaiting high office and the coalition, while planning their withdrawal, should doubt this at their peril, and I and many other Royal Marines can confirm this from our very first clinical forays into the back streets of Aden so long ago. We may have many fine professionals in uniform but there are many more out there against us who wear nothing much more than rags. As Gerry Adams has so many times warned us of the I.R.A. – "They haven't gone away you know!"

Watch my tracer on the emerging Third World and remember you read it here!

THE SPECIAL BOAT SERVICE

British Special Forces have always been recognised as the best in the world. Unlike the S.A.S., the S.B.S. have not received the publicity due to them with very good reason – secrecy. The reluctance of the M.O.D. even to admit to Special Forces has kept them in low profile and I am very glad to say that the books written by former members of S.B.S. are so few you could probably count them on the fingers of one hand. The few publications which exist were penned very carefully by men who allowed little of great importance to seep out. On the other hand as we see the bookshelves bulging to bursting point with S.A.S. publications I will simply maintain a polite and well-mannered silence. It is now of neglected consequence. I will admit to reading "Bravo Two Zero", after which I switched my attentions to military non-fiction!

The S.B.S. and S.A.S. each have proud histories fighting behind enemy lines. Their quality makes them the most cost-effective forces in the world. Recently a re-designed cap badge was introduced to replace the usual Marines badge the S.B.S. wore for many years. They changed their "Not by Strength by Guile" motto to a shorter "By Strength and Guile". Some were of the opinion that this makeover was part of an attempt by Royal Navy seniors to match the S.A.S.'s "Who Dares Wins" image.

The new S.B.S. logo shows a dagger facing upwards, with two blue lines through it portraying ocean waves.

The S.A.S. cap badge has a downward dagger with wings to underline their role in parachuting behind enemy lines.

Since the early 1990's, the S.B.S. and the S.A.S. have been under joint control of the same Director of Special Forces. Both units do similar tasks and work together on some missions. But it is a never-ending source of frustration for the highly trained and hard working members to find the S.A.S. almost always gets recognition and has, in fact, been credited with S.B.S. operations. The problem is they seldom get the kind of credit the S.A.S. always get. Yet in my experience over twenty two years of association I feel the S.B.S. are even more highly trained and I think a great credit is due to their specialist art of being, and staying, in the shadows. There is, you might say, "something of the night" about them in their invisibility, but above all their lack of interest in promotion to celebrity status.

The S.B.S. have built up a formidable reputation as a ruthless and daring unit. They specialise in counter-terrorism, beach reconnaissance, sabotage and oil platform and large ship assault. The S.B.S. was created in 1940. It became the Special Boat Squadron after the Second World War and the Special Boat Service in the 1980's.

Recently, one S.B.S. Commanding Officer wrote on the rationale for changing the S.B.S.'s regimental identity and I think his explanation was directed at every Royal Marine when he said, "As I read through the S.B.S.'s operational history just before taking command, I came across an article from the November 1947 edition of the Globe and Laurel which was written by the Commanding Officer of the Combined Operations Beach and Boat Section (COBBS) – one of the Service's many forebear units. A sentence towards the end of the article caught my eye, 'What we need at the moment is a little less secrecy so the remainder of the Corps realises what we are doing and why we are doing it.' Some things never change!"

"Mindful of this unhelpful dynamic in S.B.S.-Corps relations, I would like to take this opportunity to set out the rationale underpinning the forthcoming changes to the S.B.S.'s regimental identity.

The new SBS beret badge

From the outset, I cannot state strongly enough that changing the S.B.S.'s cap badge and associated accoutrements has nothing to do with a desire to distance the Service from the Corps. We are, therefore, very grateful that the Commandant General Royal Marines has allowed the Service to retain the Green Beret as an unambiguous statement that Commando and Royal Marine ethos will remain pre-eminent within the S.B.S."

"So why the change? The reasons are threefold. Firstly, in any organisation's, or indeed individual's, evolution there comes a time when establishing one's own identity becomes a vital prerequisite in sustaining ambition, sense of purpose and self-esteem. You do not need me to tell you that nowhere is such a dynamic so keenly felt as in the military unit, where pride in the regimental identity is synonymous with an enduring will to fight and win. As with the Corps, the S.B.S. is no different in this regard. It has a need to evolve its identity and terms and conditions of service in step with its place in the world as a unique Special Forces unit."

"The second reason is to achieve the commonality of appearance that enhances unit cohesion. Practicably, therefore, with a tri-Service approach to recruitment, there is a need to consider what cap badge successful non-Royal Marines S.B.S. personnel should wear – a similar conundrum that led to the adoption of the Green Beret by Commando Forces in October 1942. I am sure you would agree that having a myriad of R.N., Army and R.A.F. cap badges on the Green Beret would do little to shore up the cohesion of a unit such as the S.B.S."

"Thirdly, there is the issue of parity with other Special Forces Units in terms of an individual being able to take pride in having completed one of the most arduous military selection courses in the world. Whilst some may argue that seeking visible recognition for achievement in the form of a new regimental identity does not chime well with the inherent humility expected of an S.F. soldier, I would suggest that humility and taking pride in the uniqueness of one's achievement should not be seen as mutually exclusive. Again, here too there are some parallels with the logic underpinning the Corps' practices of wearing Royal Marine Commando shoulder flashes, the Commando Dagger and Parachute Wings."

"I hope that this will explain the rationale behind these changes and emphasise the importance the Service attaches to its affiliation with the wider Corps. I would finally like to thank you all for the crucial and enduring support that the Royal Marine family gives the S.B.S. There can be no doubt that the Corps remains the life-blood of the Service."

Well, I think the above explanation of on-going changes within the organisation makes it quite clear where we stand now in the early years of this new century and the position of the S.B.S. within the Corps. "But what of the S.B.S. involvement in Iraq?" you may ask. At this point in time in 2004, there is nothing I wish to say on such operations. There are, I know, a myriad of cowboys from newspapers and magazines creeping around Iraq in their desert khaki's clutching notebooks and pens, claiming to sit frequently at the dining tables and operational desks of senior coalition Commanders and coming up with stories so outlandish that they would leave the very best of Hollywood moguls out in the cold. Exactly who they offer these stories to or who, even in their wildest dreams, they expect to print or believe them is just beyond me. I will admit that some of the so-called 'true' accounts of actions that were supposed to have taken place would be wonderful material for fiction writers, but that's not what we're about here so I will stay with what I believe.

Because so little has come out on S.B.S. operations in Iraq and not to become embroiled in a world of fantasy or untruths, I will now finish our chapter on the S.B.S. by clearing up one particular story which was shrouded in confusion from reports in the early stages of the war. I want to make it plain I am only taking this line because the following tale was cleared by senior Ministry of Defence officials and nails the lie that S.B.S. troops withdrew to avoid an engagement with the Iraqi forces.

The source report on the truth of the above action deems that at the beginning of April 2003, 60 men from the S.B.S. flew into the desert of North Western Iraq. There was frustration over the failure to find weapons of mass destruction. Turkey had not allowed the U.S. to use its bases. The coalition needed a way into the North of Iraq so the S.B.S. got the most daring and dangerous mission of the war.

The S.B.S. were miles from any coalition troops, behind enemy lines and without any air support. Every other Special Forces unit in the war had planes on call very quickly.

Specially-equipped Land Rovers were flown in and the S.B.S. men used language skills and experience in working alongside local tribes to find out where the W.M.D. could be hidden.

After a couple of days driving through the desert the men liaised with the friendly Nomadic tribes and were given intelligence reports that 250 Al Fedayeen were on their way to attack them. The next night the men were in a lying up point (L.U.P.) where they used this chance to eat and rest before continuing on their task, when several streams of vehicles were spotted approaching.

Around 90 of the Al Fedayeen dismounted and started to attack. An unbelievable amount of fire opened up directly into the S.B.S. positions from many enemy points in the distance. Rounds from Duscha heavy-calibre weapons and anti-aircraft guns raked us, they reported.

The ground disintegrated around the Land Rovers as the rounds landed, the ground shook and the area smelled thick from gunfire and dust that filled the air. At first the initial reaction was one of "Where are they firing from?" but soon the guys realised that it was from an unbelievable number of guns to the front.

We estimated some 17 to 20 Duscha heavy-calibre machine guns, with at least two anti-aircraft guns. Many vehicles took direct hits but the guys continued to return fire. Tracer fire flew everywhere. The skies looked like the 4th of July. And so started a fierce firefight with the S.B.S. taking out several enemy vehicles. Mortar rounds were landing in and around our vehicles, with some taking direct hits. People were crawling along the ground to avoid being shot in order to get back to their vehicles.

The sky was alive with the sound of Iraqi heavy weapons and mortar rounds and in the distance more of the Al Fedayeen were on their way to attack from vehicles. Our commander then decided that as the mission was to discover W.M.D. and not to take on the Al Fedayeen, the unit should withdraw from the firefight and reassess with a view to possibly ambushing its attackers. The firefight had lasted some 20 minutes and despite control of the skies by coalition forces, no aircraft pitched up to help out.

"Assessing the Situation"

As it now stood, the unit, unlike any other forces attacked in the whole war, were out there entirely on their own. Some of the vehicles stayed back to provide covering fire for withdrawal as the men began pulling out. The Iraqis were gaining on the S.B. unit and after a fighting withdrawal the men reorganised themselves, and the commander decided to vacate the whole area.

The rear vehicles were engaging in a fierce fight to slow down the enemy advance. The unit began its journey out of the area using stealth and the help of night vision goggles to get away. Whenever the Iraqis got too close, the guys opened up on them to inflict as much damage as possible. Then the lead vehicles drove into a boggy area and ground to a halt. These vehicles became stuck in the bog and were unrecoverable. Just in the near distance rows of headlights lit up the skies. The Al Fedayeen hunting forces were closing in. Some of the vehicles had made it across the bog and engaged the enemy.

At this point, half of the S.B. unit were in a large bowl of land, their vehicles trapped, and the decision was taken to blow the vehicles up to prevent equipment getting into Iraqi hands. As the vehicles were being prepared for demolition the Al Fedayeen were closing in rapidly. The sky was lit up by the sight of headlights seeking out their prey. The resulting explosions attracted the Al Fedayeen. They lined the top of the bowl and opened up, thinking we were there. All the guys could see was an awesome hailstorm of yellow, red and gold streaks as tracer rounds filled the skies, going in every direction.

This gave them the vital few minutes to put a bit of distance between them and probably saved them. The guys from the destroyed vehicles climbed on to what others were left, and the group started its escape towards Syria. The Al Fedayeen formed a big ring around our men and closed in. Wherever our guys looked, all they could see was Iraqi headlights homing in on them. An overwhelming force was giving chase. Avoiding capture remained the only goal. All we had to rely on was darkness and stealth. The chasing Al Fedayeen gained on us, at times appearing to be only 100 metres away. But each time, using evasive driving techniques, the guys got away.

At one point it was decided to break into smaller groups to make it harder for the chasing force to catch anyone. Just then, two guys on one motorbike decided to make the split. The guys, an officer and a senior N.C.O., shouted, "See you in Syria," and were gone.

After nearly two hours of fighting off the enemy force, two American A-10 tank-busting jets turned up. They only had limited time there and were tasked to help one of the smaller groups to fight off the enemy. The S.B. unit was now fragmented across the desert, fighting for its survival, harassing the Al Fedayeen before moving on, giving the enemy no chance to gain the upper hand. Eventually, AC-130 Spectre gunships turned up. The Al Fedayeen were terrified of them and backed off, preferring to lie in wait and try to ambush the S.B.S.

With Spectres as lookouts, the unit made its way to be rescued by helicopters. Tired, but feeling frustrated, the men were flown out of the country and later redeployed back into Iraq to carry on their work. Back at base, the men were extremely disappointed not to have finished their mission. But this was, to date, the biggest single attack by the Iraqis on any coalition force, no matter how big. The S.A.S. had withdrawn from being attacked by a lot smaller forces and had not stood and fought. But they had the luck not to run into such a large force. Also, they had air support a lot closer to them.

As for the S.B.S., we knew it was risky, far more than usual – but that's what we are there for. The S.B.S. harbour no bitterness – just a desire to get on with the next job.

The reports made by some newspapers soon after this led me to believe that some of their writers obviously know just a little bit about an awful lot, but an awful lot about nothing. How can any element of Special Forces ever hope to make them understand the problems of identifying and avoiding strategic risks in the early stages of trailbreaking into enemy terrain.

Do I have to say it yet again? "Events, dear boy, events" – they can be as catastrophic on the battlefield as they can be in normal everyday life.

A LINE ON THE WATER -
"THE SHATT-AL-ARAB INCIDENT"

The eight British servicemen stood blindfolded in the ditch – each man lost in personal thought as he waited for a bullet to the back of his head.

They could hear their Iranian captors walking along the single line, their AK 47 rifles ready to fire. But it was just 'mind games' being played out by the Revolutionary Guards and just as they braced themselves for the first shots the six Royal Marines and two Royal Naval Ratings were marched back to their minibus to continue their transfer between jails.

This harrowing mock execution was the lowest point in their three-day ordeal after they were seized by Iranian soldiers in June of 2004. Marine Dave Reid remembered without doubt it was their worst moment. Every one of them thought they were dead. With no idea of where they were, they were blindfolded and all the guards had guns. Taken to this shallow grave, each one was convinced it was the end.

They stood there with one hand on their heads and the other one on the shoulder of the man in front so they could all walk in a straight line.

There was just ten seconds of silence, ten seconds of hell, as they waited for the bullets – and there is no doubt at all that that is what the Iranians wanted them to think. The dreaded mind games. They say the best trick the Devil ever pulled was to convince us he didn't exist. Well, in Warfighting the Devil is never far from the combat zone. The Geneva Convention and the rules of engagement are only for fools and horses, as the saying goes.

Today the men are all back home after top-level talks involving the then Foreign Secretary Jack Straw brought the diplomatic incident to an end. Iran insisted the men were seized because they had strayed into the country's waters. But then, they would wouldn't they?

The men had set off from the Iraqi port of Umm Qasr at 4.30 a.m. on Monday June 21st 2004. They were in three craft – two combat support boats or C.S.B.'s, used by engineers for tasks such as bridge-building – and a rigid-hulled craft called a Boston Whaler. The unit, under the control of Sergeant Tam Harkins, was transporting one of the C.S.B.'s to Basra along the 120-mile Shatt-Al-Arab Waterway which separates Iraq and Iran. Ownership of this mile-wide stretch of water sparked the eight-year war between the two countries in the 1980's. The border between them now runs down the middle of the channel. This fractious borderline, exactly like the landline that separates the North and South of Ireland, is a multi-million pound smuggling route for oil and livestock. The British had been training Iraqi river police, who spend most of their time tracking the smugglers.

Their day at this point was unremarkable as the hot morning sun beat down on them. Armed with just their SA80 rifles, pistols and one Minimi machine gun, they made their way up the channel, keeping their eye out for smugglers and other suspicious craft. Using maps, global positioning system devices and local knowledge, they made sure they constantly stayed on the Iraqi side of the border.

But shortly after 10 a.m. their routine mission went drastically wrong. Let's call it Murphy's Law again. It always comes into play in situations such as this!

Tam could see a couple of boats approaching in the distance and assumed they were the boats which were coming down to meet them from Basra. The plan was that he and one of the boats would go to Basra and the other lads would return to Umm Qasr. But as these boats got closer it quickly became apparent that it was not who they were expecting. There were two boats with about eight men in each. The Marines could see the craft were not marked up as military but the men in them had uniforms on and were heavily armed. One of them was manning an anti-aircraft gun and was pointing it right at the Marines' craft. Tam Harkins pointed at his charts to try to show them where they were going but their reaction was to cock and make ready their weapons.

The Marines and R.N. Ratings knew then that it was serious. Although I have said before the Gods of War hate those who hesitate, here diplomacy had to be given at least a chance.

Then, out of the blue water heat, two speedboats with three men in each, all armed with AK47's, came up from behind. The Marines thought about fighting their way out of the situation but they were hopelessly outgunned, and Tam, a likeable and level-headed Scotsman, made sure his command remained calm. He later remarked,

"We were surrounded and totally outgunned. It would have been suicidal to try to fight our way out. The bloke on the anti-aircraft gun would have cut us all down in seconds on his own."

They tried to remain calm but were getting more apprehensive as the Iranians began forcing them across the waterway and over the border. That elusive line on the water. The men did not know it at the time, but they had fallen into the hands of the Revolutionary Guards, an elite military unit who answered only to Iran's leader, Ayatollah Ali Khamenei.

As they pulled up alongside a barge anchored in the channel, the Iranians hauled Able Seaman Chris Adams, a 24-year-old reservist, off his boat, leaped onto the craft and grabbed Tam's pistol. All the men were roughly searched and had their weapons taken from them and then pushed onto the barge. The soldiers began blindfolding them with dirty rags soaked in petrol and they quickly realised they were in a hopeless situation.

Then they were made to sit cross-legged on the deck of the barge. Marine Scott Fallon, 28, a former Scottish amateur boxing champion, remembers being on the next boat after Tam and as he pulled alongside the barge the Iranians were screaming, "Weapons, weapons!" at them. Scott was trying to tell them it was his rifle and he never, ever let it go. Well, it was worth a try but they jumped down onto the boat and took the guns anyway. Again, it would have been suicidal not to hand them over. The Iranians hauled them onto the barge where they were blindfolded and told to sit down.

Scott recalls his blindfold reeked of petrol, which worried him even more. He kept his eyes tightly shut because it would have been a nightmare to get petrol in them.

Things were getting quickly out of hand and the Iranians were not only screaming at them but at each other as well.

David Reid, from Aberdeen who, with Scott, had only arrived in Iraq the day before the kidnap, remembers he was very worried at this stage and was thinking of a way of escape, but once the blindfolds went on, that was it. He was totally disorientated and the Iranians had all the weapons so they had complete control.

The situation deteriorated rapidly from then on and every step of the way they all thought they were going to die. Their biggest fear was death by beheading. They knew it was the norm in this part of the world. Their heads were buzzing with questions. Why were they blindfolded if they were only being arrested?

Tam recalled the things they were told in training which is to make your escape attempt as early as possible. So he was thinking about that but once the blindfolds go on and they have all your weapons, it becomes very difficult.

At this part of the story it is as well to remember the six Marines and two Ratings had no idea of what it was all about. They were all thinking, "Who are these guys?" Scott later remarked that even if they had escaped they would have been in Iranian territory with no idea exactly where and nowhere to go. There was only one thing to do and that was to play their captors' game, sit it out and hope that the politicians back home were doing the right thing. In the end it proved the best tactic, and their decision proved the right one.

But there were many, many times when they didn't think there would be a happy ending. After a short time the men were told to stand up and then board a smaller craft moored alongside the barge. Some journey later they were taken ashore and into a sparse room measuring about 7 ft x 12 ft. In here they were made to sit cross-legged, four against one wall, the other four against the opposite wall. As well as Tam, Dave, Chris and Scott, the Iranians had seized Corporal Chris Monan, from Cleveland, Marine Mark Warburton, from Merseyside, and Chief Petty Officer Bob Webster, an airport fireman from Newcastle who was a Royal Navy Reservist.

The eighth member was a full-time Royal Marine from Northern Ireland, but whose identity I must omit for security reasons.

Iranian officers started questioning the men using an interpreter called Hassan. It turned out to be stuff like, what was their mission and what were they doing in Iran? As Tam was their Commander he did most of the answering and kept trying to tell them that they had not been in Iranian waters, but their captors wouldn't listen.

They had all been briefed beforehand that if something like this were to happen they would be hostages and not prisoners of war. That meant they should answer all questions, not just name, rank and number, so they answered as much as they could be reasonably expected to. The first interrogation lasted about ten minutes – but it was the first of many questioning sessions during which the men were asked about British and American activity in Iraq, quizzed about personal and family details – and then told they would be tried as spies. This was what they feared most – being put on trial as spies. Now it appeared the threat would be carried out as Sergeant Tam Harkins sat there blindfolded on the floor with the interrogator telling him that this would be the fate of him and his men.

Iranian state T.V. had already said the six Royal Marines and two Royal Naval Ratings seized at gunpoint would face trial. All of the men later agreed that that was when it got really scary, when they talked about being put up as spies. Tam admitted, "If you get done for spying you get banged up for life or executed. They said to me, 'If you do not co-operate fully you will be treated as a spy.' I was answering every question I could but they started asking me about American and British policy in Iraq."

Tam and his fellow captives have spoken out about their three-day ordeal to protect their military reputations. At the time it was claimed they were well treated but they were humiliatingly paraded on T.V. At the first jail four of them asked to go to the toilet. Marine Scott Fallon recalls: "As we walked along the guards put ideas into our minds. We could peep under our blindfolds and see their feet next to us. They constantly cocked their guns. This was done to worry us – and it worked. Nearby there were people firing and we couldn't help thinking that they were practising for our execution."

Corporal Chris Monan added: "This was a clear tactic to mess with your mind. We thought a bullet was coming any second."

After 12 hours, two guards armed with AK47's entered their room, took the men's watches and tied their hands behind their backs with strips of sacking. They also used a fresh tactic – splitting the group up. Tam and C.P.O. Bob Webster were taken to another room – leaving their colleagues to worry that they had been killed.

Shortly after this they were all handcuffed with plasticuffs and taken on a two-hour journey to a second prison. Here the communal cell they were put into had no toilet or beds of any sort. The heat was unbearable and there was a bucket of dirty water, which they dared not drink.

That night they were all taken to a room with beds and air conditioning but after two hours they were returned to the old cell. This again was a classic tactic to mess with their minds and keep them feeling insecure and disorientated.

The next day they were taken by minibus to a ditch where the harrowing mock execution – which I described at the beginning – took place.

Finally, at the third prison, beds awaited all of them and they also had their first decent food. When they had eaten, a T.V. crew arrived. They felt quite angry at this, but thought it was at least a sign of freedom ahead.

Later that night came a surprise visit. Two British Embassy men arrived to reveal the Government had been working round the clock to secure their release. Someone once said that he who loses wealth loses much; but he who loses courage loses all. Amen to that.

The eight men spent a sleepless night as they looked forward to their release the next day.

CHAPTER 27

LOOKING BACK AT SPECIAL FORCES

As I search here for diplomacy while talking about Special Forces which, incidentally, people of military kind should never do, I find it taboo to speak of recent or on-going events. To do so would be enveloped in selfish indiscretion if not stupidity. Therefore, to tell you stories of their 'specialness' as it were, I feel it would not be damaging to speak of events long over and no longer of any great consequence. To do this and at the same time keep events within the Royal Marine theatre I will probably have to commit the cardinal sin for any storyteller and go back to the peripheries of a previous few chapters on the Falklands War. But not to worry! Soldiers do this all the time around corner tables in the N.A.F.F.I. so that's one good reason, and another is the almost certain prediction that for the Silver Jubilee remembrance of that war due shortly there will probably be a whole myriad of untold events being brought to surface for the first time. So, throughout the following events if you come across names and places which seem familiar let them float by and just concentrate on the actions of the real hunter class, or if you like, those with 'something of the night about them'! The Special Forces.

It seemed fitting that the glacier-strewn island of South Georgia – where the Argentine flag was first hoisted – should be the Task Force's first target. It would soon be the first place to take the Argentine flag down.

Just before noon on 3rd April Lt. Keith Mills and his 22 Royal Marines defending South Georgia ended their brave resistance and surrendered to the Argentines. On 7th April, Admiral Sir John Fieldhouse was ordered to plan the island's recapture. It was the logical first step in the controlled escalation of force that was to culminate in the removal of all Argentines from Britain's South Atlantic dependencies.

Paraquet is a somewhat archaic way of spelling 'parakeet', but that is how it was entered in the list of unused and therefore available code names for operations. This was the name selected for the repossession of South Georgia, although it was changed unofficially almost immediately to 'Operation Paraquat'.

The small Argentine force on the island was commanded by a naval engineering officer of rather dubious reputation called Lt. Commander Alfredo Astiz. There was no way in which the forces under him could properly occupy and control the whole of the wasteland of the island. They had to be content with keeping small garrisons at Leith and Grytviken. This meant that some members of the British Antarctic Survey (B.A.S.) and the two women – Cindy Buxton and Annie Price – who were making a 'Survival' film for I.T.V. remained at large.

A military presence was also kept in the South Atlantic after the surrender at Grytviken in the shape of the ice patrol vessel H.M.S. Endurance. The ship had wanted to join the fight, and was on her way to finish off the Argentine corvette A.R.A. Guerrico, when she was ordered to stand some 60 miles out to sea and keep a low profile. She remained in the area monitoring Argentine communications and sending intelligence back to the United Kingdom. She also maintained contact with the two women and members of the B.A.S. Endurance was able to stay out of the Argentines' way and eventually rendezvoused with Task Force 317 on 12th April far to the north.

The troops earmarked for the recapture of the island were from the Special Air Service Regiment, the Special Boat Squadron and the Royal Marines. Mountain and Boat Troops of 'D' Squadron 22 S.A.S. and 2 S.B.S. – a total of some 60 men – were tasked, along with M Company 42 Commando Royal Marines, and the force was under the tactical command of Major Guy Sheridan R.M., second in command of 42 Commando.

When 42 Commando left their barracks at Bickleigh just outside Plymouth on 8th April to sail down to the Falklands on board S.S. Canberra, M Company stayed behind, hiding away in the gymnasium. They were forbidden to phone their wives or families and two days later, on 10th April, they flew from R.A.F. Lyneham to Ascension Island where they joined the troops from D Sqn. 22 S.A.S. and 2 S.B.S. They all embarked on the destroyer H.M.S. Antrim and the Royal Fleet Auxiliary Tidespring. Also embarked were two Naval Gun-Fire Support (N.G.S.) parties on Antrim. The S.A.S. and S.B.S., together with some of the force command element, sailed in Antrim, and the rest of M Company were in Tidespring.

The British S.S.N. H.M.S. Conqueror was ordered to patrol off South Georgia to prevent any Argentine reinforcement of the island. The Commander of the Task Group allocated to the recapture was Antrim's captain, Captain Brian Young D.S.O.R.N. The Task Group consisted of H.M.S. Antrim, H.M.S. Plymouth and R.F.A. Tidespring. On 12th April they made a rendezvous with H.M.S. Endurance and some members of the B.A.S. team on board. The scientists briefed the troops on the Argentines' strengths and locations.

"Pre-Strike Recce"

On 14th April, Major Sheridan was ordered to plan covert reconnaissance of the Leith and Grytviken areas to determine the exact strengths and dispositions of the Argentine troops. This operation would be carried out by patrols from the S.A.S. and S.B.S. It was felt that the primary objective would be Grytviken, and Leith would come later. An initial photographic reconnaissance of the island was carried out by an R.A.F. Handley-Page Victor bomber from Ascension.

The small force sailed steadily south. As it became clear that part of Task Force 317 was heading for South Georgia, and that Britain was obviously going to take the much-speculated-about 'South Georgia option', the Argentines sent a strong platoon (some 40 Marines) to the island in one of their two ex-U.S. Navy Guppy Class submarines, Santa Fé. She managed to slip through the net of maritime surveillance which was mounted by R.A.F. Nimrods operating from Ascension Island and backed up by Victor K2 tanker aircraft. This operation was carried out between 20th and 25th April to give early warning of hostile naval movements, and swept the area from South Georgia to the Argentine mainland. Before Santa Fé could land her troops the British were in South Georgian waters.

The plan for Phase One of the operation was for 15 men of Mountain Troop D Sqn. 22 S.A.S., who were now in Endurance, to land by helicopter on the Fortuna Glacier to the north of Leith, and move by way of Husvik and Stromness to Leith itself. At the same time, 2 S.B.S. on board Antrim would land by helicopter or Gemini inflatable assault boat in Hound Bay to the south east and make their way gradually via Moraine Fjord to Grytviken. The O.C. (Officer Commanding) M Company 42 Commando, Captain Chris Nunn R.M., was to form a Quick-Reaction Force (QRF) to land as and where required. Most of M Company were on Tidespring. The B.A.S. scientists were not at all happy about putting men onto the Fortuna Glacier in the conditions that were likely to be experienced there at that time of year.

"First Enemy – The Weather"

A reconnaissance was made of the glacier at first light on 21st April by the radar-equipped Wessex H.A.S. 3 Helicopter from H.M.S. Antrim and, although there was some wind and driving rain, conditions seemed unsuitable for the operation. The Wessex 3 returned to pick up four S.A.S. men, and the Wessex 5s from R.F.A. Tidespring landed alternately on Antrim's deck to embark more. The plan was for the Wessex 3 to lead the 5s up onto the glacier by radar. The operation had to be abandoned after they encountered thick low cloud, driving rain and snow storms in Possession Bay. After some hours the weather improved, and a second attempt was made – the helicopters climbed onto the glacier in swirling low cloud. The visibility and navigation problems were made worse by frequent driving squalls of snow and sudden changes in wind speed and direction. Nevertheless, the three helicopters reached the landing zone (LZ) and deplaned their troops and their equipment. They returned to the ships by way of Possession and Antarctic Bays to avoid being sighted by Argentine observation patrols that might have been around.

During the night of 21st April the barometer fell sharply to 960 millibars and a force 10 snowstorm, which gusted to 70 knots, blew all night. The windchill factor on the glacier was dangerously high. The wind blew away the troops' shelters, and after nearly 24 hours in the blizzard and intense cold, the Mountain Troop men – under Captain John Hamilton of the Green Howards – radioed at 11.00 on 22nd April that they had been unable to move off the glacier, and that they could not survive another 12 hours and that frost-bite cases or environmental casualties were imminent.

It was decided to extract them using the same formation as before. Conditions were much worse than the day before, with swirling low clouds and driving snowstorms sweeping across the glacier. The wind was very changeable, gusting to 70 knots and then dropping unexpectedly to ten, which caused problems of severe mechanical turbulence over the mountains. It was decided to leave the Wessex orbiting in Antarctic Bay while the Wessex 3 tried three times to get onto the glacier. The 5s landed on a spit of land to conserve fuel. In the end, all three helicopters had to return to their mother ships to refuel.

A second attempt was made immediately, and this was successful. The three helicopters climbed the glacier, sighted the smoke ignited by the troops to indicate their position and wind direction, and landed there during a welcome break in the weather. But as the S.A.S. men were being enplaned, the wind blew strongly again and whipped up the snow. One of the Tidespring's Wessex 5s, call sign Y.A., had been the first to load troops and was ready for take-off, and so the pilot decided to lift immediately. As he took off and moved forward, he seemed to lose his bearings in the 'white out' and crashed, skidding for some 50 yards and ending up on his side. The other two helicopters had now embarked their troops, so they lifted and landed next to the crashed Y.A. where they loaded its crew and soldiers. Half were taken onto the other Wessex 5, call sign Y.F., which dumped fuel to carry the extra load as did the Wessex, which had the other half.

Visibility by this time was practically zero, and the wind and snow had not abated. With the survivors on board, the Wessex 3, call sign 406, took off with Y.F. following astern and they made their way down the glacier. Some seconds later, they traversed a small ridge, Y.F. was seen to flare violently and strike the top. It rolled over onto its side and could not be contacted by radio. The overloaded 406 had to return to the ship some 30 miles away to the north. The passengers were disembarked and medical supplies and blankets were taken on board. The Wessex 3 then flew back towards the glacier, but the foul weather prevented landing. Contact was however made by radio with the crashed Y.F., and it was confirmed that there were no serious casualties.

The Wessex 3 returned to Antrim to await a break in the weather. About an hour later an opportunity presented itself, and 406 flew back to the glacier and managed to locate the survivors. They were embarked and, somewhat overloaded with 17 passengers and their kit, got back to Antrim some 35 minutes later. For these feats of daring airmanship the pilot of 406, Lt. Commander Ian Stanley R.N., was awarded the Distinguished Service Order.

Hours later another team was ready to be put ashore. This time they were to go in Gemini inflatable rubber assault boats, with their somewhat unreliable 30-kilowatt outboard motors. Using five boats, 15 men of 2 S.B.S. and Boat Troop D Sqn. 22 S.A.S. set out in three-man patrols. Almost as soon as it was launched, the first boat's engine failed and it was swept out to sea. Another suffered the same fate in the Antarctic night. One crew was picked up by helicopter. The other boat managed to make a landfall on the last piece of land before the open sea.

They waited for five days before they switched on their (Sabre) search and rescue beacon in case its signals put the operation in jeopardy. The other patrols landed successfully earlier on, just after midnight, at the north end of Sirling Valley. They reported just after midday that ice from the glacier was being blown into Cumberland East Bay and was puncturing the rubber skins of their Geminis. They were recovered during the night, and the following day were put ashore in Moraine Fjord. During a lull in the blizzard on 23rd April, Mountain Troop were landed again on the Fortuna Glacier, and proceeded with their mission.

A submarine contact was made. It was thought that it was on a submerged patrol to attack the ships of the Task Group trying to retake the island, and so they withdrew towards the edge of the Exclusion Zone to the north. They made a rendezvous with H.M.S. Brilliant which was equipped with two Sea Lynx helicopters, valuable replacements for the Wessex 5s which had been lost.

During the night of 24th April the warships returned, leaving Tidespring to continue her withdrawal outside the Exclusion Zone. An intensive anti-submarine operation was begun to locate the Argentine boat. On Sunday 25th April Antrim's Wessex 3 made a radar contact, spotting Santa Fé on the surface near Grytviken, which she was leaving having landed her reinforcements. It attacked her with depth charges, which exploded very close to her port casing. It is probable that she did not dive when she spotted the Royal Navy helicopters because she thought that she would stand a better chance on the surface than submerged against the sophisticated A.S.W. weapons that the British had.

Badly damaged, she was forced to turn and head back towards Grytviken.

Plymouth launched her Wasp and Brilliant's Sea Lynx was ordered to drop an A.S.W. torpedo if the Santa Fé dived. Antrim's Wessex opened fire with her G.P.M.G., and the Sea Lynx released the torpedo and closed to harass the submarine with its machine gun. The Wasp from Plymouth sped in from 40 nautical miles away under control of the Wessex 3 and fired an A.S. 12 missile, and the Wasps from Endurance scored hits with their A.S.12s on the Santa Fé's G.R.P. fin. These missiles did little damage as they punched straight through before exploding. The disabled submarine, leaking oil and streaming smoke, was beached alongside the jetty of the B.A.S. base at King Edward's Point.

This event had shocked and demoralised the Argentine garrison. The original recapture plan had envisaged a set-piece landing by M Company, who were at this time some 200 miles away on board Tidespring. The S.A.S. D Squadron Commander, Major Cedric Delves D.S.O., on board Antrim, urged that the opportunity presented by the shock and confusion of the attack on the submarine should be seized, and an immediate landing made.

"A Scratch Company Hurries into the Fray"

On the battlefield progress has seldom been a bargain. More often than not you have to pay for it. Endurance was still in Hound Bay, and Plymouth was unable to land a Wessex 3 helicopter on her flight deck, so a composite company was hurriedly made up from all the military personnel on board Antrim. There were 75 of them in all, being elements of Mountain and Boat Troops (S.A.S.), S.B.S., headquarters personnel, some M Company Marines, members of Mortar, Administrative Echelon and Recce sections and the ten Royal Marines of Antrim's Detachment. There were also the two N.G.S. parties.

Major Sheridan set H-hour at 14.45. At H-30 minutes (2.15 pm) a Wasp helicopter lifted a N.G.S.O., Captain Chris Brown R.A., from 148 Battery of 29 Commando Regiment Royal Artillery, to a spot where he could observe the Argentines at Grytviken. Major Sheridan and 30 S.A.S. men went ashore and set up the Tactical Headquarters at Hestesletten. The rest of the ad hoc company group followed, and a mortar position was established. While this initial landing was going on, the N.G.S.O. called down fire from the 4.5 in. guns of H.M.S. Plymouth and H.M.S. Antrim, neutralising the landing site and the slopes of Brown Mountain, where Argentine troops had been seen.

At this stage in the campaign to regain the South Atlantic Dependencies, the emphasis was on limiting casualties. So the 235-round bombardment was made as a demonstration of the superior firepower of the British, and no shells were brought down any closer than 800 yds. from identified positions. The shells fell in a controlled pattern around the Argentines; the Royal Navy could have hit them if it wished. After this the N.G.S. was 'on call'.

Major Sheridan's party made their way along the steep slopes of Brown Mountain, and by 17.00 they were within 1000 yards of King Edward's Point. He ordered the landing of more troops from Endurance and Plymouth onto Bore Valley Pass, and asked the ships of the Task Force to show themselves by sailing into Cumberland East Bay. As he was doing this, the Argentines raised three large white flags in King Edward's Point settlement.

They had been persuaded that the game was up by some S.A.S. men arriving in their position by walking through a minefield and running a Union Jack up the pole. The astounded Argentine officer in charge could only protest: "You have just run through a minefield!" A few minutes later, H.M.S. Antrim appeared around Sappho Point, and the landings at Bore Valley Pass were cancelled. Major Sheridan flew across King Edward's Cove by helicopter and the Grytviken Garrison surrendered at 17.15. The next day, 26th April, the 16 Argentines at Leith were invited to surrender by radio, but they refused. A personal visit from the S.A.S. and Royal Marines, however, convinced them that they should do so without a fight. The intimidating presence of H.M.S. Plymouth and Endurance helped with the persuasion.

Among the first troops into the settlement at King Edward's Point were the medical officer and his team from Antrim and Brilliant.

They attended to an injured sailor from Santa Fé who was seriously wounded, and had to amputate a leg. He survived his injury. The only fatality was an Argentine petty officer, who was shot dead in error while the submarine was being moved under supervision. A skeleton crew of Argentines had been on board, each member with a Royal Marine guard who had instructions to prevent the boat being scuttled. Commands were to be passed down in both Spanish and English so that they could be understood. The particular order to blow tanks reached the petty officer and his guard only in Spanish. As the Argentine sailor complied with the command, the Royal Marine thought that he was about to scuttle the boat and so he shot him. Booby traps and mines which the enemy had laid were removed by them under the watchful eyes of the British troops.

The Argentine prisoners, numbering 156 Marines and Navy personnel and 38 civilians, were segregated and kept in Shackleton House until they were shipped out to Montevideo in Uruguay on 30th April on board R.F.A. Tidespring. The Argentine Force Commander, Lt.-Commander Alfredo Astiz, signed a formal document of surrender on 26th April in the wardroom of H.M.S. Antrim in the presence of Captain Brian Young of Antrim, Captain Nick Barker C.B.E. R.N. of Endurance, Major Guy Sheridan and the S.A.S. Squadron Commander.

Major Sheridan, Arctic explorer and ski racing champion, signalled to his Commanding Officer, Lt. Colonel Nick Vaux D.S.O. R.M., on board S.S. Canberra, 'Our unit flag flies high over Grytviken'. Colonel Vaux replied, 'When we have sorted out the rest of the South Atlantic we look forward to a spectacular reunion. Did you have the right wax?'

M Company remained on South Georgia, setting up O.P.s and guarding against another Argentine attempt to capture it. The island was once again in British hands, 23 days after Lt. Keith Mills' gallant fight.

CHAPTER 28

"BY STRENGTH AND GUILE"
"WHO DARES WINS"

We all know the mottoes respectively. Between them they sum up the essence of Special Forces – doing more with less, avoiding detection and making the maximum use of minimum covert effort.

"We don't talk about that sort of thing," say Government spokesmen when asked about Britain's Special Forces. And they never have. Even more than the Royal Navy, the S.A.S. and the S.B.S. are careful of their secrets. Their lives depend on silence, whether in action against terrorists in the Gulf States or in Republican bars and clubs in Northern Ireland, and on the latter you can take my word.

Unfortunately, this brand of secrecy is a sort of stamp or stain that more often than not forces you into the shadows not only for your own sake, but also for those close to you for the rest of your life. Somehow the watchers always end up being watched. The full story of their involvement in the Falklands War gets the same monosyllabic treatment – there is no full story.

Any report involving the S.A.S. and S.B.S. must necessarily be a mixture of educated guesswork and informed speculation, with the very few hard facts made available by the Ministry of Defence, and not a few flights of imagination. Other than the S.A.S. and S.B.S., not surprisingly, there are others whose work it takes a great deal of imagination to even draw close to the truth. The Special Forces make no apology for this; the less that is known about them, the greater the surprise when they crop up in some entirely unexpected quarter of the battlefield, be it a desert wasteland or a chip shop in the Lower Falls Road – the greater their chances of success. From the successes of an operation springs their hope of survival. And they very rarely fail.

For the S.A.S. and S.B.S. the Falklands War began some time before the islands were even invaded. We now know that Prime Minister Margaret Thatcher ordered nuclear-powered submarines to be sent south on Monday 29th March 1982. We can only speculate on when the S.A.S. and S.B.S. were put on alert but, as they are trained to be the spear-head of any British military operation, it seems inconceivable that (as one source put it to me), the Commanding Officer of 22 S.A.S. had to ring the Commander of 3 Cdo. Bde., Julian Thompson, asking if he wanted the S.A.S. to come along. As any operations on the Falkland Islands and South Georgia would have to be amphibious, the S.B.S. would certainly have been put on full alert; but the S.A.S. had both mountain and boat troops who would have been invaluable in the South Atlantic – and the Mountain and Arctic Warfare (M & A.W.) Cadre of the Royal Marines would have received the call sooner rather than later due to their unique training for warfare in the most hostile conditions – thus our account of the Cadre at 'Top Malo House'.

The secrecy surrounding the Special Forces was emphasised on 24th April when men of the S.A.S. and S.B.S. were already on South Georgia; 'The Task Force,' said M.O.D. spokesman Ian McDonald, 'Has not landed anywhere.' In a sense he was quite right. Neither unit was really part of the Task Force.

Their contribution to the success of the campaign was literally immeasurable – no-one except the men themselves and their masters in Whitehall knew just how much they did. Their efficiency is best judged against the performance of their Argentine counterparts; in a straight fight at Top Malo House early in June, the M & A.W. Cadre destroyed nearly their own number of enemy Special Forces for the loss of only two wounded. The reason for their success? Blind indifference to the appalling weather conditions which had kept the Argentine sentries, who should have been on watch outside, inside the house. Whatever the standard of their training they had yet to learn that the world of conflict has a funny habit of throwing up surprises when discipline and vigilance are discarded for self-interest whereas the Cadre took the pain, and in the end, the gain. For me, as I have always had a close association with the M & A.W. element of the Core, this M & A.W. Cadre action was the stuff of legend.

One of the main roles of Special Forces is to gather strategic intelligence; that is, information which the commanders and their planning staffs need to know about the strengths, dispositions, locations, morale and movements of the enemy so as to be able to launch a successful attack. In the conflict to regain the Falkland Islands and South Georgia, British Special Forces were fully employed in this way, as well as in carrying out fighting patrols to harass the enemy. The units of British Special Forces which were used were D Squadron, 22 S.A.S. Regiment supplemented with elements of G Squadron commanded by Major Cedric Delves D.S.O., No. 2 Special Boat Squadron Royal Marines commanded by Major J. J. Thompson O.B..E., R.M., and the Mountain and Arctic Warfare Cadre commanded by Captain Rod Boswell R.M.

"The Intelligence Factor"

Collecting intelligence is normally done by inserting a four-man patrol by parachute, helicopter, canoe or Gemini-type inflatable rubber boat. We are now talking of a war-fighting theatre. The patrol approaches the objective overland, and establishes itself in a hide from which it can observe the target. A short distance back from the hide would be the patrol base, where most of the soldiers rest and guard the all-important escape route. They take it in turns to man the observation position, from where they watch and make notes on the enemy. During the day they use binoculars, telescopes or cameras with long-distance lenses, and at night image-intensifying or thermal-imaging night observation devices are employed.

The information gathered in this way is normally radioed back on a high-frequency (H.F.) or satellite communication (SATCOM) link. In the initial stages of the Falklands recapture, the SATCOM sets were not available and so the patrols had either to use the H.F. link or be extracted to give their reports.

As we now know from earlier chapters on the Falklands our good friend and saviour on many occasions Major Ewen Southby-Tailyour (E.S.T.) had spent a lot of time sailing around the scattered bays and creeks of the islands during a previous posting to N.P. 8901, and had built up an impressive knowledge of those which were suitable for an amphibious landing. He was appointed to the staff of the

Land Forces Commander, and was able to suggest several valuable options.

The S.B.S. were tasked to reconnoitre in detail the beaches and shore defences of certain of the better possibilities, and the S.A.S. were required to provide information about Argentine troops in the hinterland away from the optional beaches. This was to gauge the reaction time before a counter-attack could be delivered against the beach-head, and the strength in which it would come. In choosing a site for landing it was thought that it should be a safe anchorage sheltered from the prevailing westerly winds, sufficiently landlocked to make missile attack impossible, and giving some protection from conventional air attack by free fall bombs. It would also need a safe approach from the sea. It was felt that San Carlos Water offered all these, but had the disadvantages of easily mined approaches and, with West Falkland Island being so close, naval radars and weapon systems would be degraded. Nevertheless, it was the best available landing site. However, other possibilities – such as Cow Bay – were reconnoitred by the S.B.S., who also had the job of looking for mines and beaches with obstacles, and of marking the routes in for the landing craft to their particular disembarkation points on their chosen beaches.

For these preliminary recces: the typical means of insertion for the S.B.S. was by submarine and for the S.A.S. by helicopter, although both units used these methods depending on the task and conditions. Later the helicopter became the principle mode of transport. The patrols were extracted by helicopter. Nearly all of these missions were carried out at night, and the pilots of the 'choppers' flew with the aid of passive night goggles (P.N.G.'s). These are image-intensifying goggles in a head harness, which enable the wearer to see things at their real size (XL magnification) in low light levels, such as starlight. This type of reconnaissance is fraught with problems – it's one thing to stalk artillery pieces, quite another to identify enemy units and their strengths and locations in the middle of the night without being detected. As one S.A.S. officer said caustically, 'We can't just go up to them and call the roll.' Nevertheless, the Special Forces were instrumental in sending back high-grade intelligence to the brigade and unit commanders. Inevitably, mistakes were made – but not too many. Certainly, the British Forces were far better informed about what lay ahead of them than the Argentines were.

To help them in their work, the S.A.S. had an amazing amount of special equipment. It was rumoured that they had their own procurement budget and used this to buy equipment not seen even in the British Army. Observers recall seeing U.S. Air Force transport jets unloading equipment for the S.A.S. at Ascension Island, including the Stinger SAM. Stinger is a light, shoulder-launched missile in current service with the U.S. Marines; the S.A.S. chose it for its lightness and size. It was used to shoot down one Argentine aircraft, but had no advantage over the British Blowpipe other than its portability.

Another special gadget was an attachment to the H.F. radio sets. Looking a little like a small computer terminal, it was used to encode and digitise messages; when the message was ready for transmission the terminal would send off the signal as a short 'burp' of morse or speech which would be received and unscrambled back at Brigade H.Q. However, enough of that and back to business. Well, the story goes on to say that from mid-morning on 21st May Argentine air attacks were mounted against the ships in San Carlos Water and in Falkland Sound where they were forming a 'gun line'. It was reconed that an Argentine F.A.C. (Forward Air Controller) had an O.P. on West Falkland from where he was able to direct enemy aircraft onto their targets on the gun line. A Special Forces patrol was put ashore to locate and 'neutralise' him which, it is assumed, it duly did as the enemy's selection of targets lost some effectiveness afterwards.

Special Forces patrols were sent to various parts of East Falkland as the time approached to move out from the beach-head. The priorities for reconnaissance, especially by the specialists, became the locating of Argentine 155 mm and 105 mm artillery pieces; finding enemy Special Forces (Buzo Tactico) who might attack the Brigade Maintenance Area (B.M.A.) at Ajax Bay or the shipping; locating Argentine major units; reconnoitring for minefields with the Royal Engineers; and finding out how many enemy helicopters remained intact.

On the basis of information received, Special Forces fighting patrols or Harrier air strikes were despatched. At least one Argentine artillery piece was destroyed by a laser-guided bomb directed onto the target by a laser target designator manned by an S.A.S. patrol. As the British troops moved out from the perimeter of the beach-head in a long pincer-like approach on Port Stanley, Special Forces became involved in their own fights with the Argentines. S.A.S. troops were flown forward on 24th May to Mount Kent and established themselves on the mountain (and on Mount Challenger to the south) and harassed aggressively the depleted Argentine garrison; most of them had been redeployed to the Darwin–Goose Green area to face 2 Para's attack, on the assumption that the British axis of advance would not be west-east but south west-north east. Seven days later the mountain was occupied by K Company of 42 Commando, and the S.A.S. were withdrawn.

Earlier, the Mountain and Arctic Warfare Cadre of the Royal Marines had had, as described earlier, their own private fight with the Argentines. An M & A.W. O.P. had seen the men of the Argentine 602 Marine Commando Company operating from Top Malo House, a remote building in the low land to the far south of Teal Inlet Settlement. Harriers were not available for a Fighter Ground Attack, and so more members of the Cadre were flown in and a devastating attack was launched on the house. The M & A.W. troops suffered two wounded, but killed three Argentines and wounded seven more. They took six prisoners. Because of the cold, there had been no sentries posted outside the building. One of the captured Argentine officers revealed that he was married to an English girl and he had been on military courses in Britain. This produced the irritated professional's reply, "Why the ***** didn't you follow some of the lessons then!"

Some of the best patrol reports came from the M & A.W. Cadre O.P.s. One of these O.P.s related with some mirth the sight of 3 Para advancing on Mt. Longdon and being attacked by two Pucara aircraft. As the Argentine planes hove into sight, the Paras opened fire with everything they had and the Pucaras fled from this weight of firepower. What amused the O.P. was the sight of a Para shooting at a Pucara with an 84 mm Carl Gustav anti-tank weapon. This was reported as the first 'surface-to-air 84 ever'.

Several of the statements put out by the Ministry of Defence in London during this period simply said laconically that 'aggressive patrolling continues'. During this time the Special Forces were harassing and observing the enemy, sending back reports about the effectiveness of the British artillery and naval gunfire bombardments and information about how they were causing a deterioration of Argentine morale.

Sadly, it was during this later phase that the officer commanding the Mountain Troop of D Squadron 22 S.A.S., Captain Gavin John Hamilton, was killed. He was awarded the posthumous Military Cross for a single-handed attack against an Argentine position on West Falkland. An Argentine Colonel who witnessed the action described him as 'the most courageous man I have ever seen'.

Well, the Argentine defence collapsed after the capture of Wireless Ridge (2 Para), Tumbledown Mountain (Scots Guards) and Mt. William (1/7 DEOGR) 1 and surrender talks began. The Special Forces even participated at this stage. But as always, people had to die on the battlefield for progress in this theatre has never been a bargain – you have to pay for it and so many did. True to form, the Special Forces carried the burden of int. collection and collation by and large by their training and professionalism and by never, except by chance, underestimating the enemy.

"A Detailed Look at the Action of
Captain Gavin John Hamilton M.C. S.A.S."

"Perfect S.A.S. officer material" was how one of John Hamilton's mates described him. He might have added that Hamilton's gallantry was "typical of the highest standards of the British Army officer – but that tribute came from an Argentine Colonel.

As officer commanding the Mountain Troop of 22 S.A.S., the 29 year old Green Howard was an expert climber with long experience of S.A.S.-type operations. On 21st April, he and his S.A.S. patrol were fighting for their lives on the Fortuna Glacier, west of Grytviken, South Georgia. His presence there, despite warnings from British Antarctic Survey scientists (who had briefed the Special Forces on H.M.S. Endurance), was typical of the man and the regiment; if none expected an attack from that quarter, then that's where the attack should come from. "Who dares wins" is the rule.

But it all went wrong. The men nearly died of exposure, they had to be evacuated by helicopter. In fact, Hamilton had the rare distinction of surviving two helicopter crashes in a matter of hours. What happened next is a matter of some conjecture. Men of the S.B.S. and the S.A.S. Boat Troop went ashore in Gemini inflatables just before one a.m. on 23rd April – and that same day the Mountain Troop tried again. This time it all went right. Moving off the Fortuna Glacier, Hamilton's men had an unlikely encounter with the two young British naturalists (mentioned previously as working for T.V.) Cindy Buxton and Annie Price, who were eking out a lonely existence in a hut some distance from Grytviken. For the next 24 hours the S.B.S. and the Royal Navy held centre stage, crippling the Argentine submarine Sante Fé, then landing a force by helicopter near Grytviken. Hamilton resurfaced at this point in the most spectacular way; he led his patrol through a minefield, past the astonished Argentines, and ran the Union Jack up Grytviken's flagpole.

On the night of 14-15th May the S.A.S. carried out an almost classical raid on Pebble Island, the type of operation for which they were raised by David Stirling in 1941. Some 40 S.A.S. troops were landed by helicopter near the Argentine airfield on Pebble Island, and Hamilton personally led the small team who infiltrated the airfield to destroy 11 aircraft on the ground. They did it in fine style – by placing explosive charges in the aircraft, then withdrawing under covering fire. John Hamilton's personal tally was seven planes.

The rest of his war followed much the same pattern. Five days after Pebble Island, he took part in the overwhelmingly successful diversionary attack on Port Darwin. When the main body of Task Force troops landed at San Carlos the day after this raid, the Argentines still believed that the main attack would come at Darwin. A week later he was on Mount Kent, enduring bitter weather and Argentine shell-fire as he waited for the Marines of 42 Commando to come up and relieve the small force who had taken the hill.

On 10th June Hamilton was on a hill overlooking Port Howard on West Falkland, checking on reports that the Argentines were reinforcing the garrison there. With just one man, a signaller, alongside him, he was detected by the Argentines who tried to surround the couple. In the shooting that followed, Hamilton was wounded in the back. Both men knew he had no way of escape. To give his signaller some chance of getting out of the fire-fight alive, Hamilton ordered him to withdraw while he stayed and held off the advancing enemy.

As the signaller crept away, Hamilton stood up and charged. He was cut down by a hail of bullets. He stood up again – and again he was shot down. He stood up five times, went down five times. The fifth time he didn't get up.

His signaller escaped.

Much later, when the Argentine, a Colonel, was being interrogated on West Falkland, he made a statement to his captors: "I wish to commend the S.A.S. Captain for the highest military honour. He was the most courageous man I have ever seen."

"HEROES AND LEADERS"

As we talk about the quality of men we need to sustain the ranks of Special Forces, let us just diversify again from the main subject of this book, which is the Royal Marines, for a few moments to look at the problems our training establishments are up against increasingly in this brave new century. Our reminiscing here is particularly pertinent to our "leaders" in business and political life.

I think everybody who wishes to have and maintain a successful future must have respect and regard for the past. Or if you like, any of mankind who ignores history and sacrifice neglects their own future. Towards the end of Kings Squad training in the mid 60s at C.T.C. Lympstone in Devon when one exhaustive day just emerged into the next, month after month, the few people around me who were still intact had by then learned that one of the greatest of all principles within the realms of realistic expectation was that men can do what they think they can do. It was a brave world to be part of. In a sense we were not soldiers or servicemen, but "members" of some strange force to be reckoned with. We had become creative powerhouses with a sense of purpose and duty and absolutely committed to a dignified goal.

The Victorian sage Thomas Carlyle observed that: "The history of the world is but the biography of great men." If so – and the more you think about it today, the more true it becomes – then I think we are about to run out of history. Today our leaders, whether in politics, business, sport or any other walk of life, are becoming steadily, and depressingly, less great. To be fair we have had past statesmen who were less than pure – Churchill's alcohol consumption would have killed some men. We have had adulterous ones, too

– Lloyd-George conducted Britain's part in the Great War, while running some prodigious extra-marital liaisons. The "Iron Duke" of Wellington, when threatened during his political career with the revelation of a morally disobliging story, simply retorted: "Publish and be damned." The difference was that these were big men, of great achievements and experience, who saw their foibles as petty in the context of their lives.

The accomplishments of truly great men allow them to win the respect of those they govern – and they have the self-confidence to know it. When they make mistakes – as Churchill repeatedly did in his early career – they own up, brush themselves down and get on with life. Greatness implies the strength of character to do that, and to dare any critics to do better.

Where principle and force of personality used to propel people to the top, now they arrive there by keeping their noses clean, their views uncontroversial and their cronies sweet. Too many elevated positions are filled by unreliable people who have not been tested, and who lack backbone and vision. The result in this first decade of this new century is increasingly a world without heroes, without examples to follow, and without any means of inspiration.

As with so many things, the rot starts with the political leaders. Gone are the days when they wrote their own speeches, developed their own views and took on audiences that were not hand-picked. Few go to Westminster with an ideal for public service, after having distinguished themselves in other walks of life. Now, they are mostly former public employees or party functionaries who do what they are told. Few understand the real world because few have been part of it. To complement our weak politicians, we have a culture of overpaid, whingeing senior managers. There are also overpaid, whingeing sportsmen who, when they lose, do so at the shortcomings of others. We are awash with so-called "artists" producing junk who feel they are owed a heavily-subsidised living.

There is a template of predictability in almost every trade into which any potential participant has to fit. Individuality is frowned upon and distrusted; conformity, caution and dullness are at a premium. Why are our sagging bookshelves crammed with dull, unadventurous pap, the likes of which even if bloodied would not entice a starving fox into a henhouse? Characters of instant recognition is what I am talking about. Where are today's hugely selling Oscar Wilde's with their dodgy reputations? Where are the George Bernard Shaw's in their knickerbockers? Where are the James Joyce's and the Brendan Behan's with their reek of whisky and outrageously odd behaviour?

So as we look around us in this new 21st century, what do we have? Well, not a lot, because few dare to risk rocking the boat. Risk-taking is viewed as perverse; and in any case, few are remotely equipped either by training or temperament to engage in it. Originals need not apply.

What do we mean by greatness? Fundamentally, we mean an ability to enrich the lives of others by achievement, and to change the world for the better. Leadership, originality, physical and moral courage are all part of the cocktail. The great are usually, but not always or should I say inevitably, famous. But today's cult of celebrity, so often tawdry, is not to be confused with the truly heroic. Too often we confuse fame with greatness, which is perhaps partly where we have gone wrong. Our decline of greatness has many causes. The media are partly to blame, manipulating images and personalities so that true nobility and distinction are sidelined in favour of the flashy and the superficial. The colourful and the outrageous – whose colour and propensity to outrage are often symptoms of their great qualities – have to run for cover as soon as some killjoy with a newspaper column dares to point out that they are different.

Churchill, Lloyd-George and Wellington would not have lasted long today.

There are, though, other fundamental reasons for today's great shortage of heroes. Since the last war, generations of children have been brought up in increasingly risk-free environments. It seems to have become the duty of the state and of individuals to close down any possibility of danger, excitement or adventure. Remember for example, the howls of indignation when Princes William and Harry abseiled, as schoolboys, down the wall of a dam in Wales, when the self-righteous squealed with rage about the violation of near-sacred health and safety regulations.

Britain's decline as a power has inevitably limited the outlook and attitudes of its people. There is no Empire now to defend. A hundred and some years ago Winston Churchill was at Omdurman with Kitchener's army, having just participated in the rout of 60,000 Dervishes. He was just 23 and his taste for adventure and daring, while it would bring him much unhappiness and long periods of personal doubt and failure, would in the end mark him out as the only man with the guts and political abilities to lead Britain in its hour of greatest peril. Churchill was typical of those generations before universal welfare and caution became the first priorities of the British people. Every family had its stories of recklessness and individuality.

There are men today whose fathers ran away at the age of 16 and, lying about their age, joined the army to fight the Germans. When demobilised at around 21, they would have spent years on the front line, literally up to their necks in death and destruction. Today at that age (through no particular fault of their own), the greatest risk usually taken by young men is over the sort of designer drug to take in their latest act of self gratification.

Millions of Britons in the first half of the 20th century, when barely out of school, were called upon, if necessary, to expend their lives for an ideal far higher than themselves. Death was not some great unknown; in growing up they had often seen it, with brothers or sisters dying in infancy, before the advent of advanced health care and penicillin. Childhood was not the time of cosseting it has now become. Schools were hard; discipline was strict; games were played ruthlessly. For those not fortunate enough to pass through the public or grammar school system, work started as early as 14, and was often hard and manual. It meant, thought, that people were tougher in mind and in body than they tend to be now. In that imperial age, those who wished to make their fortune were often better off going abroad. The colonies, though rife with disease and plagued by minor wars, also offered great commercial opportunities.

Most young men had been prepared at school or in the Army for the hard experiences of the colonies. From early on, they were blessed with a great sense of independence. This was a generation who knew how to survive. They had common sense in abundance, and they knew when and how to take risks. As a result, they inspired the nation and enriched themselves.

The people who thrived in such a society had to be pretty formidable. In the later part of the 20th century as their generations have died, we have seen their names splashed across the obituary columns of the newspapers. We have seen the passing of many men who distinguished themselves in war such as the R.A.F. hero "Laddie" Lucas, or Lord Lovat, who led his men at Arnhem.

Gone, too, are a generation of sportsmen who combined high living with high achievement, and whose public reputations were never less than lustrous; men such as Denis Compton and Tommy Lawton.

There were men, too, who made careers out of taking gambles that miraculously paid off, but who did so on the basis of a hard understanding of how business, and the world, worked. The easy ride was not what that generation expected. Now it is often the only ride. The present generation of sportsmen, for example – many of whom are only able to compete thanks to a regular supply of painkillers and the constant attentions of the physiotherapist – are viewed with amazement by people of the Fred Trueman or Stanley Matthews generation. These were men toughened by outdoor labour in all weathers, or worked down a mine, for whom playing even hard professional sport was not so strenuous as their earlier lives had been. Now, the culture of trying to ensure that everyone (however academically ungifted) stays in full-time education until the age of 18 – and in many cases 22 – completes the process of insulation from reality.

Can we hope for greatness in a world that has done all it can to abolish hardship, effort and risk? Can anyone who has not had to struggle, to suffer in some way, or to take great gambles, have the strength of character and the force of personality to make a great impact on our world and our lives?

The answer is, probably, no. But then the cynics would argue that in a world that has been so sanitised of all unpleasantness as ours has, and in which anything remotely resembling pain, effort or grief have been eliminated, we may not need great men in the way we used to in the more dangerous, less predictable times.

Well, as the Duke of Wellington also said: "If you'll believe that, you'll believe anything."

To my reckoning, too many generations now have lived with no real war and no sacrifice. To date the U.N. has witnessed 250 wars (of sorts) since its beginning. But on many occasions it was not aided through unhelpful public attitudes. Today it's all about rights and nothing about duty. People at some point very soon must come to realise that if they value comfort and money more than vigilance and duty they will lose them. If they value peace above freedom they will lose freedom. If a country is hapless enough to allow its pacifist self-serving yuppies and Pharisees to put on show, day after day, their claims on human rights and political correctness, then mark my words, that country will lose out.

In the first decade of this new century we are fighting an idea, a form of political Islam that sees the West and everything we stand for as corrupt, degenerate and contemptible. How do we defeat them? The first thing is to recognise we are fighting a disease. We must quarantine and isolate every cell. We must plant informers within every gathering of what the world of intelligence considers fanatics and to do this we need covert people of the hunter class. We don't want readers of the health and safety bible. We don't want worriers or depressants who don't realise how cold it can get when they're out there alone with the enemy. So, what does that leave us? It leaves us in need of real heroes and leaders, not Rambo's, but real class professionals of the old school. To find these people it is necessary to search deep into the ranks of the military, for to find them outside of that form of discipline is so rare in today's values it would more than likely be written off as nigh on impossible.

To get back to our original subject what we are talking of here is the quality of the Special Forces and the history of the service they have given to this country.

Many great soldiers have emerged from our Home County regiments. Some regiments are so obscure as to be virtually unknown to the general public. Could this be once again a neglect of history? Probably. Few knew, for example, that "Monty" had a huge association with the Royal Warwickshire Regiment as most only knew of him as the Army Commander. Unfortunately the same negative attitude holds sway today even with the famous front-line die-hards like the Royal Marines, Para's, Engineers, Artillery, S.A.S. and S.B.S. All are deemed as wonderful until there is the slightest hiccup, usually instigated by the failure of the sitting government of the day, and then the infamous rumours hit the newspapers with the force of a princess dealing with an unwanted pregnancy.

Earlier I tried to put the record straight on the S.B.S. being reported in the papers as having "run" away from the Al Fedayeen in Iraq and now, once again, I find it painfully obvious because of weak and ill-informed press reports around January 2004 that I must put another record straight on "vast" numbers of Royal Marines being evacuated from Winter Deployment (WD '04) from Norway.

OUTRAGEOUS FORTUNE
OR LIES, DAMNED LIES AND STATISTICS!

Those of you who read and watched the U.K. national media during the later part of February 2004 could be forgiven for thinking that the Royal Marines had turned into a bunch of ill equipped weekenders who couldn't keep warm in the cold and were always falling off their skis.

"Old sleeping bags blamed for Marines' frostbite", screamed the headline from one broadsheet, above an article penned in the warmth of a U.K. office and part-based on reports filed by regional media. Coincidentally, my internationally known source of information used in bringing to surface the true facts and cutting the untruths was a completely independent observation team who attended all the briefings in Norway from where these horror stories originated. They also ensured they had C.D. or printed copies of all the presentations to refer back to for accuracy but the story they heard was not the same one that the national newspapers printed and aired just a few days later.

Anyone who has even a remote knowledge of the Royal Marines or who has served with the military will understand the concept of training hard to fight easy. I believe from long experience this is one of the first maxims any British soldier learns, and I say it is the cornerstone of military instructor training. Why else would our armed forces train so much in such inclement places like Dartmoor, Sennybridge, Otterburn and Salisbury Plain at the very times of the year when the weather is the soldier's worst enemy? It is only by training in conditions that are as bad as you are likely to face on the potential battlefields of the world that you can hope to come out on top.

The typical, and much-quoted example of this as we have well covered now was the 1982 victory in the Falklands over terrain that was deemed too difficult to cross and in weather conditions that left the unprepared enemy shivering in their trenches. True, many of the troops who participated in the epic fight against both nature and the Argentinians were well-prepared Royal Marines, but their frontline companions included Paras who had the primary role of fighting light in temperate or warm environments, Guardsmen more used to ceremonial duties in London, and Gurkhas trained primarily for jungle fighting. It was not the quality of their equipment, which in most cases was inferior to that of the enemy they faced, but the high standard of their training, that saw them succeed where even Britain's friends thought they were likely to fail.

The Royal Marines and British Army soldiers of 3 Commando Brigade have been the U.K's winter warfare specialists for decades and are classed as elite forces by many of Britain's friends and enemies alike. As mentioned in earlier chapters the ordinary Marine endures around nine months of intensive training on the longest infantry course in the world before he earns the right to wear the dark green Commando beret, and even the previously fully-trained Army gunners, engineers, logisticians and mechanics who support the Royal Marines infantry units have to survive weeks of body punishing Commando training before they can don their green berets and deploy with the Brigade. However, and here we come to facts and figures of paramount importance which the media seem blissfully unaware of, over the years of 1996 to 2004 elements of the Brigade had been operationally deployed so many times that their specialist winter training had to be severely curtailed. From April 2004, the last time that 3 Commando Brigade undertook a full Brigade field training exercise was actually way back in 1996, when they deployed to North Carolina, instead of Norway, for the massive Purple Star joint exercise with the old 5 Airborne Brigade and the U.S. Marine Corps. The last Brigade exercise in Norway back in 1998 was considerably under strength due to operational commitments elsewhere, and the Arctic exercise of 2001 was severely curtailed by the outbreak of foot and mouth disease in the U.K. This resulted in seventy per cent of the Brigade being Arctic novices, and around ninety per cent of 40 Commando were ski and snow survival newcomers at the start of the exercises of 2003-2004. To compound matters, these were the guys who were pulled out of Norway two-thirds of the way through the exercise to prepare for operational deployment back again to

Iraq to play the hearts and minds game little more than a year after going ashore there in a combat role; talk about stretching capabilities! That kind of multi-tasking would have been a real front-page story for the nationals to have got their teeth into! But no. Only derisive news was good enough.

The exercise was called Winter Deployment 2004 (WD 04) and the aim was to deploy a force of 2,400 personnel, including two Commando units with artillery, engineer and full logistic support, by sea and air into an Arctic environment and then train them to survive and operate in the severe terrain and hostile climatic environment to be found there, at minimal cost to the U.K. taxpayer; the word 'train' being the important one in the sentence, though it must be admitted that 'minimal cost' also played a big part. At the end of the nine-week training phase, a little under half of the troops were scheduled to return to the U.K. for the specialist preparations required for operations in Iraq, where +30C temperatures could be expected instead of the –30C of Arctic Norway, but the remainder of the Brigade was to participate in a multi-national amphibious assault about three hundred kilometres inside the Arctic Circle as part of the week-long exercise Joint Winter '04.

For the average Commando, be he infantryman, gunner, engineer or logistician, the WD 04 package commenced with two weeks of novice and continuation training, consisting of initial lectures, basic skiing practice and survival skills tuition. This was immediately followed by the week-long infantry Winter Warfare Course, where the basic tactics and fieldcraft needed to cope with the challenging Arctic terrain and climate were taught. One good illustration of just how different conditions are for soldiering in a Norwegian winter is the water situation. As full water bottles would quickly freeze solid on the belt or in the Bergen, resulting in them splitting at the seams or forcing off their caps due to the expansion, troops simply do not carry them in the Arctic. There are a few exceptions but on the whole they are considered a waste of space on a heavily laden soldier. However, despite the external cold, the type of exertion needed to battle through deep snow carrying your ammo, food and home on your back, while still maintaining a comfortable body temperature, means that dehydration is always a problem and fluid intake has to be watched as closely as if you were in the desert.

Despite being constantly surrounded by water, albeit in frozen form, getting sufficient to drink is a major problem and very time consuming. It is not until you actually try to heat snow on a portable stove that you realise just how long it takes to thaw. Remember that basic science lesson at school where you heated a beaker of water with a couple of ice cubes in it and watched the thermometer sit at zero until the last cube melted? Now multiply that by a hundred and you will see why it can take two hours to melt down enough snow to give sufficient water to re-hydrate rations and provide sufficient drinks to replace water loss after a four hour yomp or ski patrol. It is only after physically taking soldiers into this hostile operating environment and teaching them basics such as this in a realistic scenario, that you can prepare them to fight rather than just survive. The down side of this is that a small percentage will fall by the wayside and become training casualties, but better that than becoming an operational casualty through being insufficiently prepared.

Every Royal Marine deployed on WD 04 was sufficiently proficient with a rifle to be classed as a marksman in many other armies, as are the vast majority of the Commando-trained Army personnel providing support to the infantry. Artillery crews, mortar teams, machine gunners and snipers are also highly trained in their specialist skills. However, the adverse Arctic climatic conditions have an impact on all weapons systems and their performance, which means that different care regimes, operational procedures and tactics must be learned for the Commandos to perform as well in the Arctic as they did in the equally challenging high and dry Afghan mountains or the hot and humid Al Faw Peninsula in Iraq. To ensure everyone was up to speed with their Arctic drills, week 4 of the course concentrated on live-firing and mobility under extreme conditions.

During week 5 of WD 04, when the component units went into their company level exercises to validate the effectiveness of individual soldiers, sub-units and equipment in a field training exercise (FTX) environment, they were closely watched by our independent team of observers. Monday was taken up by travelling to the port town of Harstad, near to the Allied Training Centre (North) at Asegarden where the Brigade H.Q. was based for this phase of training. First thing next morning they headed off up north on the two and a half hour road journey to the H.Q. of 42 Commando at Bardu for a briefing with its O.C. Lieutenant-Colonel "Buster" Howes. This was followed by another ninety-minute road move north-east on packed ice to see the Manoeuvre Support Group live-firing their weapons. It is only when you are actually on the ground in Norway that you realise just how vast this country is; at around 1100 kilometres from Oslo, the range they were on is almost as far from the Norwegian capital as London is. The observers who spoke to the guys on the firing line were soon to learn that even simple tasks like switching machine-gun barrels has to be done differently out there. For example, if you lay the hot barrel on the ground after the swap, it will just plunge through three feet of snow, or worse, with a big hiss and will be near impossible to find if you don't stop firing immediately to look for it. The steam generated in some conditions by rapid fire also has to be catered for, both to be certain you are still hitting the right target in poor visibility, and to ensure your position is not marked for the enemy by your own personal cloud hovering above.

By Thursday of that week the observation teams were watching second echelon troops undertaking their week-long survival training package. They had already spent three nights living in ten-man tents and learning the basics of survival, such as how to build various snow structures for protection, including snow holes called 'Quincys' that can accommodate half of an infantry section in reasonable comfort. Constructed by simply shovelling snow into a hard-packed mound and then hollowing it out to give a basic igloo with an internal sleeping shelf to keep the occupants out of the cold, these snow holes protect against the worst of the Arctic conditions. With a single candle both providing light and heating the air, the temperature inside stays just above freezing, which is warm enough for the human body to survive if the right clothing is worn and proper precautions are taken. Living in a snow tomb may not seem like much fun, but those taking part in these training programs soon learn that so long as the soldier listens to his training teams and is properly equipped, it is actually reasonably bearable.

Talking of training and equipment, now is the time to bring up those sleeping bag and frostbite stories that the national media went to town on. In addition to seeing troops out in the field, the independent team of observers which included a representative from Navy News monthly, were given many briefings from Captains through to

Brigadier on the Commando side, plus one by the Norwegian Colonel in charge of the Allied Training Centre (North) as well as being granted a comparatively informal interview with Norway's Minister of Defence, who they had earlier met while she was visiting the Brigade's Tactical Command Post. At one of the briefings, the Norwegian Colonel mentioned that it was mainly from past experience of working with the Royal Marines that Norway now had what he considered to be the best issue combat uniforms and sleeping bags among N.A.T.O. forces. By comparison, the current British sleeping bag, as at WD 2004, which was conceived in response to lessons learned in the Falklands, did rely on yesterday's technology, but this was already being addressed by an extra liner being introduced, though as the combination weighed over five kilos the search then went out for a lighter and more high-tech replacement. The current British sleeping bag of WD 2004 was designed as a reasonable all-rounder for all theatres that British forces could expect to be deployed to, rather than being a dedicated extreme cold weather item for issue to an army that could be operating for half the year or more in Arctic conditions in its national defence role, so it could not reasonably be expected to match the Norwegian one.

One area where we could improve the sleeping bag issue system, and for that matter the short-term issue of many other special-to theatre items of clothing and equipment, and which could do with a bit of a shake-up is the method of tracking usage of individual articles. At present, if a sleeping bag, to take just the topical example, is relatively well cared for and is well laundered before being put back on the stores shelf, there is no way of telling whether it has merely survived one mild winter exercise on Salisbury Plain, or has been well used on three six-month operational tours and half a dozen three-month Arctic exercises during its lifetime. This very point was discussed with the media by a logistics officer in a Q. & A. session after his briefing, although he had already told them that steps were being taken to introduce a traceability system that would ensure the history of every sleeping bag would be known to ensure that the quartermaster had more than just a visual test to go on.

Bottom line was, however, if the sleeping bag of WD 2004 and the breathable bivvy bag were used properly inside a well-constructed snow hole or a heated tent, there would be little possibility of any soldier going down with frostbite as the national media were gleefully claiming. During the many observer briefings, everybody from Brigadier downwards was perfectly open about both casualty rates on the exercise and the lack of previous experience that seventy per cent of the personnel deployed had of Arctic conditions. It was even freely admitted that 40 Commando had, as at WD 04, a ninety per cent novice rate, as a result of them being to the fore on active operations rather than training in the snow, for as long as most ordinary soldiers and junior N.C.O.'s had been in the Corps. The whole point of taking 3 Commando Brigade up into the Arctic Circle was to ensure that the relevant skills were learned and relearned in case they should be needed in a future conflict, and if I may dare to look back at this point in our story and say that on the very first "Clockwork" exercise some thirty five years ago when the core was charged with the defence, in part, of the northern flanks of N.A.T.O. and under the command of the relentless Captain Mike McMullen I was very quick to learn the now sound adage:

It is only by training hard that a soldier can learn to fight easy.

Thinking back to those days of 45 Commando, the new arrivals at Condor in Arbroath, I think it fair to say that Mike McMullen made sound foundations for what has always been one of the most challenging and unpredictable theatres, logistically, for any soldier to operate in.

One by-product of training hard today is that casualties will always be taken. However, not only will casualty rates be lower if the troops are properly trained like this before going into an operational environment, but the attrition rate during training is bound to be higher as peacetime rules apply. So, let's get a few figures straight here for the record. The following figures were freely given to the independent observers in black and white. After four weeks of novice, fieldcraft and live-firing training, including over half of the Brigade learning to ski for the first time and more than two out of three undertaking a week-long survival course in temperatures that fluctuated between +5C and –35C, total casualties were under five per cent. If you are the sort of person whose cup is half empty that equates to 112 personnel, but if it's half full you would realise that a more than respectable ninety five per cent learned all the lessons and were lucky enough not to fall by the wayside. Of that total 57, or just over half, were affected by frostbite or frost nip, 22 suffered from trench foot caused by the unseasonably mild conditions and ensuing thaw that affected Norway on WD 04 part-way through what should have been the extreme cold dry survival phase, and 53 had orthopaedic injuries caused primarily by falling off their skis in training or taking a tumble on sheet ice. It is very difficult for the lay person to be aware of how treacherous sheet ice on a gentle slope can be when it refreezes after rainfall and a partial thaw. I was surprised to hear that the casualty rates were not higher. From memory, I can remember during the early years of Winter Warfare training casualty rates of just 3 per cent when the Brigade annually deployed to Norway, so 5 per cent is not bad in my books for deep snow newcomers. You cannot make an omelette without breaking eggs.

Sleeping bags and casualty rates made easy headlines for those national media journalists sitting comfortably in their warm offices two thousand kilometres away, but they did not give the full story. In addition to bringing everybody up to speed on ski and survival training, the Brigade was also trialling two types of brand new skis that will hopefully provide better mobility for heavily laden troops and reduce orthopaedic injuries caused by the traditional Telemark ski design which gives little ankle protection. They were also working through the implications of the new four-company Commando 21 doctrine under Arctic operating conditions, and experimenting with BV 206 troop carriers to see how the imminent introduction of the armoured Viking over-snow vehicle would affect tactics. However, only bad news sells newspapers, the nationals are not interested in good news stories, so all the hard work of the troops and the relentless patience and skill of the long-suffering training instructors never get a look in. The newspapers look for the worst possible scenario and promptly give it the "publish and be damned" treatment. If they really wanted a bad news story here's a couple they could try for size. They could have homed in on the effects of over-stretch on long-serving members of the Brigade, or even highlighted how budget restrictions prevented one officer from taking a sizeable portion of his command on this much-needed exercise, but that does not make such a good headline as "Marines get frostbite in the Arctic". It's an old adage but it's as true today as it ever was; the first victims of journalistic incompetence are investigation, confirmation and then truth. In this day and age you can never be sure of a headline that is printed with a view to selling a newspaper any more than you could, or should be absolutely sure of a dappled flickering shadow in tiger country. It could be something or nothing, or it could be the instant death of a good reputation. These lies, damned lies and statistics suffer from an absence of boundaries. Now we come to the punchline; if you happen to be a "hero" on the stage or football field you just sue for millions, on the other hand if you're a soldier in a snow-hole or a desert sanger you just keep quiet and soldier on.

CHAPTER 29

"THE IRISH – FRACTURED LOYALTIES"

Well, before we turn this book into a political platform and before I turn away from the art of bashing the duly elected Right Honourable Members of this and that, let me just briefly touch on a subject which many books in general seldom dwell on for any detailed length of time. Once again, sadly, it's probably politics! But what else could it be?

Throughout my entire career with the Army and the Royal Marines, I have met and worked with literally thousands of servicemen from both Northern Ireland and the Republic of Ireland. From our previous actions and events you have probably gleaned by now that I was born and raised in Ireland during those never-to-be-seen-again times of the 1940s-50s, when the trees were green and the world was young and the small town of Avoca had not yet been condemned to be forever more known as "Ballykissangel". God, what a legacy for an innocent simple village; however, that's another story.

I have always found within the Royal Navy and the Royal Marines the uttermost tolerance for ethnic minorities. Unfortunately this tends to slip, just a little, in the much bigger world of the Army. It takes courage to be a soldier these days, but it takes a special type of courage to be an Irish soldier in the British Army. Irrespective of the merits or otherwise of their choice, they are often bereft of the support structure soldiers take for granted. In many cases they cannot count on the support of their country family or friends, although for me this was not the case as the "Griffin" clan have historically served the Crown and that is to say, Army, Navy, Air Force, Marines and elements of Special Forces which I cannot, nor would I wish to go into. Others are not so fortunate and they would not be treated as heroes among their own people. Their choices are often made against the grain of everything political, national and historical they learned growing up.

Before we continue, allow me to explain that I am not pontificating here in any way, but when word got out that this book was on the cards and people of Irish origin who are now serving within the Royal Marines, some in Special Forces, became aware that being Irish myself and I would be describing events in Northern Ireland, many members contacted me and asked me to bring to light some of the facts now that their greatest patron, the Queen Mother, is no longer here to visit and speak to them. The war in Iraq during the early part of 2003 brought to the fore the Irish presence within the British Army and in particular the colourful emergence of Lt. Col. Tim Collins, the Commander of the 1st Battalion, the Royal Irish Regiment.

The melancholic dilemma of being an Irish serviceman in the British Army, made more poignant still by the death of Lance Corporal Ian Malone of the Royal Irish, serving in Iraq in April '03, was summed up in W. B. Yeats' famous poem "An Irish Airman Foresees His Death".

"Those that I kill I do not hate,
Those that I guard I do not love."

The folk memory and deep distrust of the British Army is ingrained deep in the Irish psyche. The massacres at Drogheda and Waterford by Cromwell's soldiers, the execution of the leaders of the 1916 Rising, the behaviour of the Black and Tans, Bloody Sunday and so on – the history of Ireland as we understand it has been a struggle against and not for the British Army. And yet, nothing sums up the schizophrenia, collective amnesia and downright hypocrisy of Irish attitudes to Britain more than the attitudes to Irish soldiers serving in the British Army.

From the ending of the Penal Laws, many more Irishmen fought for rather than against the Crown. Irish troops led by the Irish-born Duke of Wellington were prominent at the Battle of Waterloo. By the 1860s, half the British Army – then at the apotheosis of its imperial conquest – were Irish, prompting the old saying: "The British Empire was won by the Irish, administered by the Scots and Welsh, and the profits went to England."

More than 3,500 Irishmen died in the Crimean War, 500 alone in the week of the Easter Rising died on the Western Front and 49,000 in all died in the First World War.

More extraordinary still, in the Second World War 120,000 servicemen and women in the British Army gave their next of kin as in officially neutral Eire – proportionally more than came from Loyal Ulster. Eighty years, and counting, after independence from Britain, the tradition continues as personified in the death of Lance-Corporal Malone from Ballyfermot in Dublin. He was one of 30 southern-born Irish soldiers serving with the 1st Battalion Irish Guards in Iraq. By all accounts, the Micks, as they are affectionately known, acquitted themselves splendidly. As part of the 7th Armoured Brigade their taking of the city of Basra was a model of urban warfare with few casualties.

Just like their predecessors, the desire to serve Queen and country among Irish soldiers is always likely to be a secondary consideration. Some join because of unemployment at home. Ranger Alan Boyle of the Royal Irish Regiment said he joined because there was a three-year waiting list for the Irish Army.

"There's a lot of people out there who just want to soldier. They are not bothered which Army they are with as long as they can soldier," he said.

Others join out of a sense of adventure and because they want to experience the sting of battle. Lance-Corporal Malone joined because he grew tired of working in warehouses and packing plants. For some insane reason he was deemed too old at the age of 22 to join the Irish Army. Before he joined the British Army he had never been out of Ireland. When he was interviewed for an RTE, Irish T.V. documentary "All the Queen's Men" the year before in 2002 he had visited 20 countries.

This well-made documentary was, and is, a testament to a man at ease with the obvious contradictions of being an Irish soldier in the British Army. He laughed at the bawdy images of Irish Guards singing along to "The Merry Plowboy" and such Irish songs in their barracks, while later standing to attention for "God Save the Queen" on the parade ground. Nor, for him, were there any contradictions in swearing an oath of allegiance to the Queen. He told the T.V. programme: "At the end of the day, you can't walk away from that.

I don't think that's an option at all. That type of disloyalty would be far worse than the disloyalty of going abroad to join the British Army." He was absolutely right!

"I'm just doing my job. People go on about Irish heroes dying for freedom, that is a fair one, but they died to give me the freedom to choose what I want to do." Again, bang on target.

The Irish Guards were formed in 1901 by Queen Victoria in recognition of the bravery of Irish troops in the Boer War. As an armoured regiment, they are frontline combat troops, but their role in Iraq was also to win over hearts and minds. It's a role which suits the regiment, said Captain Alan Gardner of the Irish Guards from Coleraine:

"We have to fight and then provide the peace support. Because we are Irish, we can do both and do it very well."

Another enduring image of the Iraq war was that of Belfast-born Lt. Col. Tim Collins, the laconic, cigar smoking, Ray-Ban sunglasses-wearing Commander of the 1st Battalion of the Royal Irish Regiment. His speech delivered on the eve of battle has been compared to Churchill and Shakespeare's famous call to arms in Henry V.

"If you harm the regiment or its history by over-enthusiasm in killing or in cowardice, know it is your family who will suffer," he told his men. "You will be shunned unless your conduct is of the highest for your deeds will follow you down through history. We will bring shame on neither our uniform or our nation."

The amalgamation of the Ulster Defence Regiment and the 1st and 2nd Battalions of the Royal Irish Rangers formed the Royal Irish Regiment around the year 1993. More than 60 per cent of the regiment were Irish born, with 20 per cent coming from the Republic. With the entire regiment deployed in the Gulf, Major Mike Ruddock estimated that there were more than 100 soldiers from the south in the Royal Irish serving in Iraq.

The entire regiment certainly conformed to a certain type of Irish stereotype.

"The only thing that will stop us getting to Baghdad is if the Iraqis place an off licence in the way," Lt. Col. Collins said on St. Patrick's Day. Both Irish regiments have acquitted themselves with distinction in the Gulf, but they are still unlikely to come home as conquering heroes. Sixty years ago, Irish soldiers who fought in the biggest cataclysm in history, the Second World War, returned to a country either indifferent or hostile to their experiences. "They didn't want to know," said former paratrooper Tony Geraghty, who parachuted into Normandy after D-Day. His glider crashed killing eight of the twelve soldiers involved, all Irishmen.

"They didn't die for King and country, they died for freedom."

The prospect of dying in a foreign field unsupported and unremembered at home is one that preys on the minds of many Irish soldiers in the British Army. Wandering through the crosses at Normandy, Corkman Sergeant John Cronin came across so many Irish names.

"They have been properly respected and recognised in France, but back home they have relatives who can't even talk about it," he said. "I often wonder if I die in a conflict, will my family be able to talk about it. I don't think they will. It's a disgrace really."

Colonel Tim Collins bowed out of the Army on April 8th 2004. He left as a bitter man and who can blame him? As Commander of the 1st Battalion, Royal Irish, he earned the admiration of the world with his stirring eve-of-battle speech, urging his troops to be ferocious yet magnanimous in victory. His reward was to find himself facing trumped-up war crimes charges.

Colonel Collins was cleared absolutely of mistreating Iraqi prisoners but he has every right to feel betrayed. The Ministry of Defence and the top brass distanced themselves from him while the allegations were being investigated. After collecting an O.B.E. from Prince Charles on his last day in uniform, Col. Collins remarked: "It is for others to reflect on whether they behaved honourably towards me."

Well, as a former Royal Marine and to compound some of this new century's injustices, let me bring to light something else that was in progress across London as Tim Collins was receiving his O.B.E. A High Court Judge was delivering yet another treacherous blow to the Armed Forces. I bring this to the fore because something quite similar happened to me with 45 Commando in Belfast in the early 70s.

Two Kosovans were being awarded, by a High Court Judge, £100,000 each because they had been fired on by paratroopers during a night of madness in Pristina in 1999. The two Kosovans were in the car, which disobeyed an order to stop, and attacked the Army patrol with rapid fire from automatic weapons. The para's saw what turned out to be a relative of the two men aim an assault rifle at them and shot him dead. The men in the car claimed they were celebrating a ceasefire, but N.A.T.O. had already warned local people not to carry guns.

What the hell were the paratroopers supposed to do? As I said above, I was in a mirror image situation in Belfast and I certainly returned fire because not to do so in my books was political correctness gone mad. The Judge deemed that because the troops were on peacekeeping duty they were not entitled to "combat immunity" and the M.O.D. was ordered to pay damages on top of the £170,000 in legal aid the two Kosovans had already received to sue the Army. What a unique madness we foster! The Government orders young soldiers to risk their lives sorting out anarchy in a faraway country about which they know virtually nothing – and then allows them to be sued for damages for defending themselves against gunmen. Then, as always, this never-ending flow of money has to come from the long-suffering British taxpayer, already funding peacekeeping operations in Kosova and elsewhere.

I continue to search for answers as to why and for how much further into the years to come will British troops continue to put themselves in mortal danger on the orders of politicians, when all they can expect back home is total betrayal and an ever-increasing political inclination towards deployment overstretch.

Is it really any wonder that real heroes and much needed senior officers like Col. Tim Collins have decided the game isn't worth the candle? Collins knew, as many more now know, that in and around 2004 Iraq would become enveloped in a cadre of opportunistic power-hungry rebels which would ensure U.S. and U.K. forces would be garrisoned in Iraq in a two-pronged war, (i) internal civil security, (ii) terror-guerrilla warfare right into the unforeseen future. It was in mid-2004 after just 9 minutes of train bombing in Madrid that Spain demonstrated to global outlawry that such acts could not only bring down the leader of a country but change the Government as well. And then to cap it all the withdrawal of Spanish forces from Iraq. This was not the way for Spain to go.

When the first bombs exploded in Northern Ireland in 1969-70, if the British Army had packed their bags and run away from the danger and responsibility, unknown to most people, the Irish Defence Forces were camped and assembled just south of the border ready to move into Northern Ireland to protect the Catholic communities from their Loyalist neighbours. Thankfully the above did not come about and if it had the Irish Army would have proved hopelessly inadequate, the border (unrecognisable in most areas) would have fractured and the whole problem spread nationwide. This did not happen because the British security forces persevered. Those old enough to remember will know, the act of turning on the radio each morning brought the same monotonous news: 'In Belfast last night, another soldier!'

It's a sad thing to say but this is exactly the sort of perseverance needed in Iraq. At time of writing this in 2004 life in Iraq and Afghanistan can only be described as utter chaos. But packing bags and walking away from lawlessness is not a luxury we can afford in this new third world war against – well, the third world!

Today in Iraq and other such theatres of operation, life is as it is. Life is never as we want it. It's how we handle it that makes the difference.

CHAPTER 30

SOME POINTS TO CHEW ON

On turning into the new century at the time of the Afghan-Iraq invasions and the period of occupation/nation-building, some of the people I approached for information, remarks and opinions on the relationship between the military and political arenas were scathing, if not at times bordering on mutinous. The rapport between the two could only be described as being similar to the one between the Captain and crew of H.M.S. Bounty. But the military should have expected this as it's not the first war in which things were not as they could and should have been.

I dare say you have by now assessed my vision of politicians, and the views of the learned people I have spoken to could not agree more. This includes one or two very old acquaintances who are now quite senior members of N.A.T.O. and who, for obvious reasons, I cannot mention by name. They understand the rapid slide politicians frequently take into oblivion. The ones who have no positive agenda or working knowledge of the job they have had plunged onto their shoulders and this is particularly relative in the strategic and complicated area of defence. Politicians are wonderful "lookers". They like to "look" at things. They are all there treading the expensively tiled and carpeted corridors of Westminster. The free-booters and opportunists, the expense-sheet manipulators and the pharisees, the hucksters, chancers and snake-oil salesmen. When and where an unwelcome problem rears its head they will engage in age-old methods of solving it, particularly where great expense is concerned, such as defence. If they suspect an argument is looming with the Admiralty for instance, the Government will promise to "ringfence" this, and "copperfasten" that, thereby creating "packages of measures" which they then "look at"! After a respectful period, should the M.O.D. or the public still have the audacity to further question any progress, then the politicians while agreeing to "look at it again", will quickly build up more rocks around the top of their protective sanger by promising to hold an "internal enquiry", which then gives them the lee-way to answer all further questions with the ultimate stonewalling tactic of "wait and see".

Now, while not being a very religious man, I think I'll get away with saying I remember from school being told it was that unfortunate man on the cross who looked up towards heaven saying: "Father, forgive them, for they know not what they do." Well, if ever it needed repeating with conviction, that need is now in the form of a huge broadside in the direction of global politics. Today there is an unparalleled absence of honesty in military related politics. Quite apart from the Arab world and its peripheries, right here at home there is the unfinished business of Northern Ireland.

I won't go into this again; suffice to remind people, as should an open Government, that the Provisional I.R.A. are still active and recruiting. They are in a high state of readiness and frequent training continues all over Ireland and on many sites in Britain. The Provo's maintain a capability in the area of intelligence and on weaponry which is continually upgraded, and senior Sinn Fein members retain considerable influence on P.I.R.A. policy. Now, don't look for this in the national press, because you won't find it, as they are probably too busy scandalising soldiers who were caught smoking in some foreign field, and you certainly need not wait to hear the above from the mouths of politicians. But believe me, on our approach to the reunification of the provinces which is a certainty in the not-too-distant future, even most Loyalists accept this as inevitable, there will be many bulls let loose in many china shops as the now covert hardliners roll out their munitions. The ghettos are simmering!

But as I keep saying, that's another story for another book.

As we move towards closure on this particular book and to move onto a lighter story, I have been asked to remind readers of an old stalwart who has had many years of association with the Royal Marines and for most of his life could be found tirelessly pounding and punishing his body on all sorts of antics for the benefit of the General Infirmary at Leeds.

When I mention the above Infirmary, you may have guessed the man concerned is Sir Jimmy Savile, O.B.E., K.C.S.G., to whom I and the few who remained in one piece in the 861 Squad in late 1967 will forever be indebted for a much needed two-hour break in training. This happened early one morning when we were only weeks away from passing into the Kings Squad at the end of nine months of hell. There was a full dress rehearsal for the presentation of the Green Beret and that, in 1967, meant fifty minutes of non-stop drills – again and again. Normally the days are not so bad, but at this particular stage of training when you are tested to the limit and beyond, any break is heaven sent.

We paraded at 07.00 in full blue uniform but nobody was looking forward to almost an hour of pounding the parade ground. The day before, in late afternoon we had completed a nine-mile speed-march in less than the allotted time, and now the next morning (as is normal) exhaustion had set in as a haze of tiredness, and heavy parade boots aggravated raw blisters. Then, (and this could only happen to me!) I looked down to discover that oil from the S.L.R. rifle had seeped from the gas regulator and onto the forefinger of my gleaming white glove.

Well, I don't know about today, but way back then when breathing could be a restricted privilege, this state of affairs – if the Adjutant saw it – could result in me cutting every blade of grass on Exmoor with a pair of scissors, and then a bit extra on Dartmoor. But the day was saved when we all caught sight of the First Drill (instructor) approaching with an out-of-place-looking civilian who was wearing an apprehensive look and clutching a clipboard.

Most of the nonchalant squad were too bleary-eyed to pay him much attention but as he came closer, and being Irish myself, I straight away recognised him as Aemonn Andrews, the T.V. "Big Red Book" man. The drill instructor then gathered us all around in a tight half circle and the plan was revealed and conspiratorially hatched.

Apparently, on numerous occasions – for reasons only known to himself – Jimmy Savile had wriggled out of being trapped onto the T.V. show but it seems they were keen to get him. And get him they did. Now, this was quite a relief for me because in those days, and being a vain sod, I was slightly worried that Andrews was coming for me. That'll be the day, says I!

"The Cunning Plan"

If you remember, it was around that time the Royal Marines were taking on part-responsibility for the northern flanks of N.A.T.O. and lots of weird and wonderful new kit was appearing on the scene, especially with 45 Commando in Stonehouse Barracks. The whole idea was to disguise Aemonn Andrews within the ranks of the Kings Squad which Jimmy Savile was asked to come and visit. But the very idea of trying to pass Aemonn Andrews off as a fully trained and dress-rehearsal-ready Royal Marine of Kings Squad standard could only be akin to trying to pass Nora Batty ('Last of the Summer Wine') off as a modern day Liz Hurley. The problem here was of Saturn rocket proportions.

To use that old expression again, the day was saved by the drill instructor, who told us to go away and get changed into field kit which was to include a rather large parka (sort of anorak) which had fur fringing all around the hood – we were to wear this in the 'up' position. The idea was that Andrews could do the same and no-one could tell the difference if he fell in in the front rank next to me. We would simply be a troop of men in Arctic clothing. This would, and it did, fool Jimmy.

The "event" was to take place at 09.30. We were into autumn, but it was still very warm and any onlookers could only have concluded that the lunatics had taken over the asylum on Commando Training Centre's parade ground. The Adjutant's horse, who it was planned to have on the original rehearsal, but was now not required, was being perambulated for exercise on the grass verges of the parade. He seemed to know he was no longer part of this and while dropping a large contribution just looked the other way in complete disgust at the whole frightful charade. Horse sense!

But it worked. It worked as only a drill instructor can make things work in outlandish situations. Aemonn stood next to me with one parka sleeve hanging loose because his arm was inside holding the enormous red book with which he would dish the dirt on Jimmy. He took it very well, for he knew he had no way of escape, and as he said to me in his letter recently – August 2003 – the Royal Marines succeed where mighty T.V. companies fail. Amen to that Jim. We certainly fixed it for him that day!

When this was all over my main objective then was to scurry away and scrub my oily glove until it gleamed again and get it dry for the main parade of the Kings Squad, which had been put back on at 13.00 hours. Looking back now at crazy days such as that one, it's easy to understand how the late great David Niven, as I think I mentioned before, found the material for his best-seller "The Moon's a Balloon". There are frequent incidents in military life that can be found in no other profession, and by this I mean outlandish situations which can only be laughed at. In my opinion, this can only be a good thing, for as time passes there is, and will be, less and less for soldiers to celebrate, for as I write this near the summer of 2004, the Marines of 40 Commando are yet again heading for deployment to Iraq to fill the gaps left by the Spanish troops who have packed their bags and come home. I do not blame the Spanish troops, nor their joy at being free of that place, for it is yet again political shenanigans.

Sir Jimmy Savile OBE KCSG

The General Infirmary at Leeds
Great George Street
Leeds LS1 3EX

Telephone 0113 3923965
Email mavis.price@leedsth.nhs.uk

Mr D Griffin
9 Rest-a-while Park
174 Ringwood Road
St Leonards
Ringwood
HANTS BH24 2NR

14 August 2003

Dear David,

The following couple of paragraphs are Jimmy's recollection of the day in question.

"This Is Your Life tried on two separate occasions to trap me. Both attempts failed so they spoke to the Royal Marines at Lympstone who said "bring your cameras *on such a date* and we will make sure that Jimmy is here.

The Royal Marines asked me to inspect a troop that were going to the Arctic. I dressed up for this and with full military honours walked down a line of these Marines who were dressed in Arctic clothing with fur anoraks round their faces. One of them was Eammon Andrews who surprised me with the book and it goes to prove that Royal Marines succeed where mighty TV companies fail"

I hope this is useful

Yours sincerely,

Mavis Price

M P Price

EPILOGUE

Tough job putting this book together. How does one capture the moment and the atmospherics in life and death situations? I have tried to set the scenes for such events as the deaths of Cpl. Leach and Michael Southern, of Sgt. McKay and Capt. Hamilton and so many others, but combining cold hard facts with dignity and spelling-it-out as it were proved daunting.

My primary purpose here was to do justice to the men and events I spoke of and the actions they took part in. My other objective was to present these stories in a "matter of fact" genre as would be related from one man to another in plain boots-on-the-ground speak. Sometimes many history books, as I have said before, can come over to the reader as a series of very dour, cold and clinical statements bordering on 'reports'. I wanted to take this book a more casual way but at the same time maintain its strictly non-fiction status. I wanted the reader to find it easy to take in, be they male, female, young, old and even those who know little or nothing of military matters. In fact, dare I say, there is nothing in these stories which would make it unsuitable for school children or the families who have lost members in any military conflict you care to mention. It's as true to the point, plain and simple as that.

The main problem with writing a history on any military theatre, I would suggest, is glaringly obvious. If you weren't there, such as events in the First and Second World Wars, then you can't write from experience. You are then left to trawl through everything you can for factual information from other books of substance to newspapers and magazines, from period publications to scraps of paper at the Ministry of Information, not to forget museums and even cemeteries.

More important are those individuals who have a working memory of the events you wish to talk about and I again acknowledge all of those organisations and people who were willing to talk to me, and who politely put up with my monumental persistence and insistence on as fine a detail as possible. There is a huge credit due to their patience and understanding.

On many occasions throughout these stories I wanted to bring sadness and suffering to the fore, and in many chapters it does not make for politically correct reading but the truth was, and is today, plain to see and I stand by it.

In the past this sort of story-telling of events of war was done by such giants of literature as the likes of Norman Mailor with "The Naked and the Dead", a novel of men and warfare, and Michael Herr with "Dispatches". Their methods of expressing conflict, its highs, lows and all of the emotions that enter into war inspired me to turn what could have been a long and drawn out record of historical events into a compact spread of events with limitations on the boring and difficult-to-understand bits. This method of story-telling will, I hope, fire the imagination of even those with the most sedentary of natures. And what better way to do this than in the company of the Royal Marines by sea, land and "air" for I dared not leave untold their periods of piloting the skies for the Royal Navy.

And finally I would suspect there are people out there who would not agree with my opinions and methods of visiting vulgar indignity onto many of our leaders and politicians. To them it may seem to offer a degree of disrespect. But, as on one or two occasions I have almost died for their freedom to give these opinions, on equal terms I think I can offer mine and those of many I have spoken to. To me this is a deliberate avenue towards a reversal of patronisation. With them I take no prisoners. I think I speak for many people today as in modern democracy the lay person is so often a stranger to the facts surrounding his lot in life. He is, in not so many words, expected to shut up and get on with it.

Churchill warned of this and along with those other men of honour, Eisenhower, Montgomery, and yes, even Rommel, I can only imagine them either turning in their graves in frustration, or perhaps (and I hope this is the case) doubled up laughing at the iniquitous nonentities who have barged their way into positions of high office in the early years of this century, particularly within the European sphere of events.

Many politicians today are unchanging people in a changing world. They are out of step, out of time and some totally out of control. I have little faith in any of the above surviving one hour of Kings Squad drill on the parade ground at Limpstone, have you? Some time ago someone described Westminster to me as so many thousands of square feet of fantasy surrounded by reality. What's your opinion? I leave that for you to decide.

And what of me, PO-2521O-H? Well, I have many years ahead of me to promote Her Majesty's Royal Marines and when I finish, no doubt another will take my place in the slit trench, for I am surely not, nor will I be left as the last man standing. As time goes by there has to be someone willing to stand up and tell of the positive and slam the negative. This is not whistle-blowing, it is simply telling it as it is and taken that the only exact science known to man is hind-sight, when I find something is very wrong nowadays – militarily that is – I will drag it to the surface and expose it, not to make waves or rock the boat but for the poor sod in the trench who has no voice on the matter.

So, this leaves me to now answer those who accuse me of the besmirchment of the "Right Hon. Gentlemen" who hover over us, or those who think I have been scathing towards politicians.

I think I can answer this perfectly by resurrecting and bringing to life once again those immortal words of Rhett Butler as he walked out the door on Scarlett O'Hara during the last scenes of "Gone With the Wind" -

"Frankly my dear, I don't give a damn!"

ACKNOWLEDGEMENTS

I would like to acknowledge all of the following people and organisations who willingly helped me with the production of this book. I'm sure most history writers will agree that no matter what pains you go through to bring forth the most accurate description of events within the genre of non-fiction warfighting, there will almost always be someone out there who will dig out and bring to light the smallest of discrepancies not for the importance of the mistake, but just for the pure hell of proving you wrong. It happens to all writers, even to those giants of the bookshelves. This is why over three years I left no stone unturned to iron out the truth.

What you read is what I found and what you don't read is what I won't tell you. Take my word, you wouldn't believe what can be found under some stones! So, from the excellent memories and writings of the following people from individuals to huge publishing organisations, I extend my thanks for the help which made this book possible. I say this not only for the richly deserved and continual memory of all those who never came home again and their families who were left wanting, but also because some of my opinions are also those of many experienced people who are as yet, unlike me, not in a position in life to openly give theirs. I have said many things that needed saying in this new century of high-octane buffoonery and political flummery.

I'm sorry to say the first person I would like to mention seems somewhat out of reach. Try as I may I have failed to contact David Barzilay, author of 'The British Army in Ulster, Volumes 1 and 2'. His publisher was Century Books, Belfast, which now no longer exists. If David reads 'Of The Hunter Class', I would ask him to contact me A.S.A.P.

The following have extended me, courteously and without question, every help I asked for:

De Agostini Rights Ltd.

International Literary Agency. For the memoirs of Ewen Southby-Tailyour in Orbis 1997 issue 114 Volume 10 of Elite Magazine

A particular thank you to the Managing Editor of Marshall Cavendish Partworks Ltd., a member of The Times Publishing Group (Singapore)

The Historical Records Office Royal Marines

The editorial staff of The Globe and Laurel magazine

The editorial staff of Soldier magazine

The editorial staff of Navy News

The Publishing Editor of Eye Spy magazine

The Publishing Editor of Combat and Survival magazine

Her Majesty's Stationery Office – Crown copyright material including The Royal Marines, The Admiralty Account of Their Achievements 1939-43 (1944)

The Imperial War Museum London

The Royal Marines Museum Eastney, Hants

The Admiralty Library

Naval Historical Branch (Naval Staff)

Ministry of Defence

The editorial staff at the Daily Mail London

The editorial staff at the Irish Post London

The Chief Reporter and War Correspondents at The Sun newspaper London

Finally may I extend my thanks to the management and staff at 'Raffles Hotel', Singapore, for putting up with my comings and goings over the past 4 years. We will meet again soon.

AUTHOR'S NOTE ON THE MISCONSTRUCTION OF HISTORY

As a matter of interest on trying to continue history on the straight and narrow and as close to the truth as possible, just let me put the following example to you.

The wording to the Preambles to the Articles of War and succeeding Acts has changed over the centuries and right up to matters of war in recent years:

See these famous words on the Royal Navy:

"It is upon the Navy,
under the good providence
of God, that the wealth,
safety and strength of the
kingdom do chiefly depend."

Most Royal Marines and Royal Naval personnel will have seen the above somewhere during their service.

Now let me give you some slight changes and anonymous alterations to this anonymous original, which have led to one of the best-known phrases in the English language being one of the most commonly misquoted:

The most recent version of the Preamble is found in the Naval Discipline Act 1957 (itself now amended by the Armed Forces Discipline Act 2000), the opening sentence of which reads: 'Whereas it is expedient to amend the law relating to the government of Her Majesty's Navy, whereon, under the good Providence of God, the wealth, safety and strength of the Kingdom so much depend.' This form of words was first used in the NDA of 1866 (29 and 30 Vic. c.109).

The 1661 Articles of War – properly the 'Act for establishing Articles and Orders for the regulating and better government of His Majesty's Navies, Ships of War and Forces by Sea' (13 Car.II.Stat. i.,c.9) – were not actually the first such; there were certainly two earlier Acts sanctioned by Parliament, of 25th December 1652 and of 5th March 1648/9, but neither had the preamble. The phrasing of the Preamble to the 1661 Act in part reads '….His Majesty's Navies, Ships of War and Forces by Sea, wherein, under the good Providence and Protection of God, the wealth, safety, and strength of this kingdom is so much concerned.' The original author of the 1661 version is unknown, but it was certainly not King Charles II.

The preamble of the amending Act of 1745 (18 Geo.II c.35) reads '….His Majesty's Navy, wherein at all times and more especially in time of war, the wealth, strength and safety of these kingdoms are so much concerned.'

The version given in the Act of 1749 (22 Geo.II.c.33), which continued in use till the 1860s is '….H.M. Navies, ships of war and forces by sea, whereon, under the good Providence of God, the wealth, safety, and strength of this kingdom chiefly depend.' The 1758 'Act for the encouragement of seamen employed in the Royal Navy' – not a Naval Discipline Act – has this passage '….the Marine Force of this Realm, whereon, under the good Providence and Protection of God, the security of these kingdoms and the support and preservation of their trade and commerce, do most immediately depend.'

The concepts within the Preamble are foreshadowed in The Libel of English Policy, a 1100 line poem of 1436, which has this:

'Kepe then the sea that is the wall of England
And then is England kept by Goddes hand'

Even earlier, the Commons in 1415 recorded that 'la dit Naveye est la griendre substance du bien, profit & prosperitee du vostre dit Roialme' (Rolls of Parliament, 3 Hen.V).

So, why is this? Is it Murphy's Law again, or a kind of "Mission Creep" in the world of literature? It's all very interesting which just goes to show and prove how the elasticity of history is a game we all but play at as we strive to cruise the peripheries of truth and its upkeep.

There is no doubt that everything written today on the Falklands, Ireland and Iraq will, by 2050, be outrageously prone to misinterpretation. But then I guess this promotes every positive reason to continue to take pen to parchment in striving to keep our line of vision on an even keel.

But don't you agree that for a positive and constructive future we should, as I have said before, follow the only realistic and true science known to man – hindsight!

David Griffin
Ringwood, Hampshire
January 2005

ISBN 1-41205463-X

9 781412 054638